Terahertz Technology and Its Applications

Terahertz Technology and Its Applications

Editor

Victor Pacheco Peña

MDPI • Basel • Beijing • Wuhan • Barcelona • Belgrade • Manchester • Tokyo • Cluj • Tianjin

Editor
Victor Pacheco Peña
Newcastle University
UK

Editorial Office
MDPI
St. Alban-Anlage 66
4052 Basel, Switzerland

This is a reprint of articles from the Special Issue published online in the open access journal *Electronics* (ISSN 2079-9292) (available at: https://www.mdpi.com/journal/electronics/special_issues/terahertz_technology).

For citation purposes, cite each article independently as indicated on the article page online and as indicated below:

LastName, A.A.; LastName, B.B.; LastName, C.C. Article Title. *Journal Name* **Year**, *Volume Number*, Page Range.

ISBN 978-3-0365-0996-9 (Hbk)
ISBN 978-3-0365-0997-6 (PDF)

Contents

About the Editor

Victor Pacheco-Peña received his PhD degree, cum laude, at Public University of Navarra (Spain). He holds a prestigious Newcastle University Research Fellowship, and he works within the School of Mathematics, Statistics and Physics at the same institution in the UK. He has published more than 100 papers in high impact journals and international conferences. He was a visiting researcher at the Imperial College London (UK) and University of Pennsylvania (USA) in 2014 and 2015, respectively. From 2016 to 2018, he was appointed as a Postdoctoral Fellow within the Department of Electrical and Systems Engineering from the University of Pennsylvania (USA). He has been awarded as "Young Scientist" by the URSI GASS 2020 in Rome, Italy, "Young Scientist of the Year" by the Spanish URSI during the XXXI Spanish Conference URSI 2016 and has received a CST University Publication Award for Best International Journal Publication using CST Microwave Studio (R) in 2016. He serves on the Editorial Board of several international journals including Frontiers in Photonics, *Applied Sciences* (MDPI) and *Electronics* (MDPI). His current research is focused on Metamaterials, THz, Plasmonics, Optics and Electromagnetism.

Editorial

Terahertz Technologies and Its Applications

Victor Pacheco-Peña

School of Mathematics, Statistics and Physics, Newcastle University, Newcastle Upon Tyne NE1 7RU, UK; victor.pacheco-pena@newcastle.ac.uk

Citation: Pacheco-Peña, V. Terahertz Technologies and Its Applications. *Electronics* **2021**, *10* , 268. https://doi.org/10.3390/electronics10030268

Academic Editor: Hirokazu Kobayashi
Received: 23 December 2020
Accepted: 21 January 2021
Published: 23 January 2021

1. Introduction

The terahertz frequency range (0.1–10) THz has demonstrated the provision of many opportunities in various fields, such as high-speed communications, biomedicine, sensing, and imaging [1–6]. Historically, this frequency range, lying between the fields of electronics and photonics, was known as the "terahertz gap" because of the lack of sources, detectors and fabrication technologies.

However, considerable effort is now being devoted worldwide to improving this technology. Within this context, great progress has been made to fill the gap in this interesting spectral range, such as multiplexers and tuneable devices [7], among others. The aim of this Special Issue is to provide a platform to highlight the work being conducted within this range of the electromagnetic spectrum.

2. In This Special Issue

This Special Issue consists of thirteen papers covering a range of applications using THz technologies, including THz sensing and imaging, spectroscopy applications, and non-destructive testing. The contents of these papers are introduced below.

Reference [8] presents the modelling and evaluation of zero-biased Schottky diodes. Two different mounting techniques are considered: wire bonding and flip-chip. The experimental results are supported by numerical simulations demonstrating the validity of the proposed models. The improvement of radar cross-section using THz signals is shown and demonstrated in Reference [9], where the concept of adaptive gates is adopted to reduce the signal-to-noise ratio, thereby improving the accuracy of the measurement. The design of a frequency multiplier source working at 0.335 THz is reported in Reference [10] with two different schemes; experimental validations of the proposed designs are provided.

This Special Issue also includes applications in THz spectroscopy. In Reference [11], the authors propose a mechanism to improve the accuracy of optical delay lines for THz spectroscopy applications by using an optical encoder. In Reference [12], the authors demonstrate a radiation power improvement of almost four times for spiral photoconductive antennas. Analyses of the structures are carried out using THz time-domain spectroscopy. In Reference [13], it is shown how THz spectroscopy can be used for non-destructive testing of the hollowing deterioration of stone relics (Yungang Grottoes in this case). Further applications for non-destructive testing using THz radiation are presented in Reference [14], where an optimal scanning technique for honeycomb sandwich composite panels is proposed. THz spectroscopy is applied in Reference [15] to evaluate the vulcanization and macrodispersion of silica for rubber products using THz absorption measurements.

The design, study, and experimental demonstration of a biased sub-harmonic mixer working at a frequency of 0.67 THz is presented in Reference [16], demonstrating a conversion loss of 18.2 dB in the band between 0.650 THz and 0.690 THz. A synthetic aperture THz imaging technique based on the light field imaging system is proposed in Reference [17]. An on-chip THz detector is presented in Reference [18]; it is designed by using both an on-chip inset-feed rectangular patch antenna and a catadioptric lens. Reference [19] presents a nano displacement sensor using hetero-structure waveguides working in the THz frequency range of 0.8–1.1 THz, demonstrating a maximum sensitivity of around 1.2 GHz/μm.

A coupled stack oscillator working at 0.350 THz is presented in Reference [20], showing an output power of −0.8 dBm at 0.3532 THz.

Funding: V.P-P. is supported by Newcastle University (Newcastle University Research Fellowship).

Acknowledgments: I would like to thank all of the researchers who submitted their work to this Special Issue. Their contribution is invaluable, and they have equally contributed to make this Special Issue a success. I would also like to express my gratitude to all of the reviewers who helped in the evaluation process of all of the manuscripts, made important suggestions, and contributed to improving the quality of the accepted manuscripts. I would also like to acknowledge the support from the Editorial Office of Electronics who worked extremely hard to maintain the high standard of the journal when the Special Issue was live and supported me during the whole process for a rigorous and timely peer review of the manuscripts.

Conflicts of Interest: The author declares no conflict of interest.

References

1. Pham, H.H.N.; Hisatake, S.; Minin, O.V.; Nagatsuma, T.; Minin, I.V. Enhancement of spatial resolution of terahertz imaging systems based on terajet generation by dielectric cube. *APL Photonics* **2017**, *2*, 056106. [CrossRef]
2. Sun, L.; Zhou, Z.; Zhong, J.; Shi, Z.; Mao, Y.; Li, H.; Cao, J.; Tao, T.H. Implantable, Degradable, Therapeutic Terahertz Metamaterial Devices. *Small* **2020**, *16*, 2000294. [CrossRef] [PubMed]
3. O'Hara, J.F.; Withayachumnankul, W.; Al-Naib, I. A review on thin-film sensing with terahertz waves. *J. Infrared Millim. Terahertz Waves* **2012**, *33*, 245–291.
4. Pacheco-Peña, V.; Engheta, N.; Kuznetsov, S.; Gentselev, A.; Beruete, M. Experimental Realization of an Epsilon-Near-Zero Graded-Index Metalens at Terahertz Frequencies. *Phys. Rev. Appl.* **2017**, *8*, 034036. [CrossRef]
5. Pacheco-Peña, V.; Beruete, M.; Minin, I.V.; Minin, O.V. Terajets produced by dielectric cuboids. *Appl. Phys. Lett.* **2014**, *105*, 084102. [CrossRef]
6. Freer, S.; Gorodetsky, A.; Navarro-Cia, M. Beam Profiling of a Commercial Lens-Assisted Terahertz Time Domain Spectrometer. *IEEE Trans. Terahertz Sci. Technol.* **2021**, *11*, 90–100. [CrossRef]
7. Karl, N.J.; Mckinney, R.W.; Monnai, Y.; Mendis, R.; Mittleman, D.M. Frequency-division multiplexing in the terahertz range using a leaky-wave antenna. *Nat. Photonics* **2015**, *9*, 717–720. [CrossRef]
8. Gutiérrez, J.; Zeljami, K.; Fernández, T.; Pascual, J.P.; Tazón, A. Accurately modeling of zero biased schottky-diodes at millimeter-wave frequencies. *Electronics* **2019**, *8*, 696. [CrossRef]
9. Pang, S.; Zeng, Y.; Yang, Q.; Deng, B.; Wang, H.; Qin, Y. Improvement in SNR by adaptive range gates for RCS measurements in the THz region. *Electronics* **2019**, *8*, 805. [CrossRef]
10. Meng, J.; Zhang, D.; Ji, G.; Yao, C.; Jiang, C.; Liu, S. Design of a 335 GHz frequency multiplier source based on two schemes. *Electronics* **2019**, *8*, 948. [CrossRef]
11. Mamrashev, A.; Minakov, F.; Maximov, L.; Nikolaev, N.; Chapovsky, P. Correction of optical delay line errors in terahertz time-domain spectroscopy. *Electronics* **2019**, *8*, 1408.
12. Kuznetsov, K.; Klochkov, A.; Leontyev, A.; Klimov, E.; Pushkarev, S.; Galiev, G.; Kitaeva, G. Improved InGaAs and InGaAs/InAlAs Photoconductive Antennas Based on (111)-Oriented Substrates. *Electronics* **2020**, *9*, 495. [CrossRef]
13. Feng, J.; Meng, T.; Lu, Y.; Ren, J.; Zhao, G.; Liu, H.; Yang, J.; Huang, R. Nondestructive Testing of Hollowing Deterioration of the Yungang Grottoes Based on THz-TDS. *Electronics* **2020**, *9*, 625. [CrossRef]
14. Im, K.-H.; Kim, S.-K.; Jung, J.-A.; Cho, Y.-T.; Woo, Y.-D.; Chiou, C.-P. NDE Terahertz Wave Techniques for Measurement of Defect Detection on Composite Panels of Honeycomb Sandwiches. *Electronics* **2020**, *9*, 1360. [CrossRef]
15. Hirakawa, Y.; Yasumoto, Y.; Gondo, T.; Sone, R.; Morichika, T.; Minato, T.; Hojo, M. Application of Terahertz Spectroscopy to Rubber Products: Evaluation of Vulcanization and Silica Macro Dispersion. *Electronics* **2020**, *9*, 669.
16. Ji, G.; Zhang, D.; Meng, J.; Liu, S.; Yao, C. Design and Measurement of a 0.67 THz Biased Sub-Harmonic Mixer. *Electronics* **2020**, *9*, 161. [CrossRef]
17. Lyu, N.; Zuo, J.; Zhao, Y.; Zhang, C. Terahertz synthetic aperture imaging with a light field imaging system. *Electronics* **2020**, *9*, 830. [CrossRef]
18. Zhao, F.; Mao, L.; Guo, W.; Xie, S.; TH Tee, C.A. On-Chip Terahertz Detector Designed with Inset-Feed Rectangular Patch Antenna and Catadioptric Lens. *Electronics* **2020**, *9*, 1049. [CrossRef]
19. Xu, L.-L.; Fan, Y.-X.; Liu, H.; Zhang, T.; Tao, Z.-Y. Terahertz Displacement Sensing Based on Interface States of Hetero-Structures. *Electronics* **2020**, *9*, 1213. [CrossRef]
20. Nguyen, T.D.; Hong, J.-P. A 350-GHz Coupled Stack Oscillator with −0.8 dBm Output Power in 65-nm Bulk CMOS Process. *Electronics* **2020**, *9*, 1214.

Article

NDE Terahertz Wave Techniques for Measurement of Defect Detection on Composite Panels of Honeycomb Sandwiches

Kwang-Hee Im [1,*], Sun-Kyu Kim [2], Jong-An Jung [3], Young-Tae Cho [4], Yong-Deuck Woo [1] and Chien-Ping Chiou [5]

1 Department of Automotive Engineering, Woosuk University, Wanju-kun, Jeonbuk-do 55338, Korea; wooyongd@woosuk.ac.kr
2 Division of Mechanical System Engineering, Jeonbuk National University, Jeonju, Jeonbuk-do 54896, Korea; sunkkim@jbnu.ac.kr
3 Department of Mechanical and Automotive Engineering, Songwon University, Gwangju 33209, Korea; jungja@songwon.ac.kr
4 Department of Basic Science, Jeonju University, Jeonju, Jeonbuk-do 55069, Korea; choyt@jj.ac.kr
5 Center for Nondestructive Evaluation, Iowa State University, Ames, IA 50011, USA; cchiou@iastate.edu
* Correspondence: khim@woosuk.ac.kr; Tel.: +82-63-290-1473

Received: 20 July 2020; Accepted: 19 August 2020; Published: 21 August 2020

Abstract: Terahertz wave (T-ray) technologies have become a popular topic in scientific research over the last two decades, and can be utilized in nondestructive evaluation (NDE) techniques. This study suggests an optimal scanning technique method for honeycomb sandwich composite panels, where skins were utilized with two different skins, namely, carbon fiber-reinforced plastic (CFRP) skin and glass fiber-reinforced plastic (GFRP) skin, as layers of the panel surfaces. Foreign objects were artificially inserted between the skins and honeycomb cells in the honeycomb sandwich composite panels. For this experiment, optimal T-ray scanning methods were performed to examine defects based on the angle between the one-ply thin fiber skin axis and the angle of the electric field (E-field) according to the amount of conductivity of the honeycomb sandwich composite panels. In order to confirm the fundamental characteristics of the terahertz waves, the refractive index values of the GFRP composites were experimentally obtained and analyzed, with the data agreeing with known solutions. Terahertz waves (T-rays) were shown to have limited penetration in honeycomb sandwich composite panels when utilized with a skin of carbon fibers. Therefore, T-rays were found to interact with the electrical conductivity and electric field direction of honeycomb sandwich composite panels with glass fiber skins. The T-ray images were obtained regardless of the electric field direction and the fiber direction. In the honeycomb sandwich composite panels with carbon fiber skins, the T-ray images with higher signal-to-noise (S/N) ratios depended on the scanning angle between the angle of the carbon fiber and the angle of the electric field. Thus, the angle of optimum detection measurement was confirmed to be 90° between the E-field and the fiber direction, particularly when using a carbon fiber skin.

Keywords: terahertz waves; honeycomb sandwiches; foreign materials; time-of-flight; electric field

1. Introduction

Recently, utilization of terahertz wave (T-ray) technologies have increased exponentially in technical applications such as mechanical aviation, aerospace, and advanced medical fields, with field practical applications also revealing broad application prospects. T-rays, which have relatively short wavelength and high resolution, are widely used in fields of inspection using electric and electronic spectra. In particular, T-rays are critically important in security devices used in airports, advanced imaging, liquids, various industrial areas, and spectroscopic evaluation of advanced composite materials [1–5]. Terahertz time–domain spectroscopy (THz-TDS) plays an important role in contactless detection of discontinuity or defects, which are present in various composite materials. The THz system is based on photoconductivity and depends on the generation of low-cycle terahertz waveforms utilizing photoconductive sensors mounted with femtosecond (10–15 s) lasers [6]. This system has the ability to generate picosecond terahertz waves and obtain a high signal-to-noise ratio (S/N). This energy affects a wide range of bandwidths and the resistance of photoconductive switches could cause a temporary change in THz wave emission to be produced over the THz timescale baseline [6–9].

The other method uses a laser of two continuous waves (CW) via optical conversion and optical mixing. Much attention surrounds T-ray signals because of their use in monitoring, such as management and inspection of nonconductive products, chemical components, and physical properties and substance toxicity analysis. Since T-rays are employed in small, portable pieces of equipment using previously discovered foundation technologies, its range of applicability is wide and the utilization of these technologies may be significant.

More recently, the importance of fiber-reinforced plastics (FRP), which are used in industrial regions as well as state-of-the-art aviation sectors, was identified. Characteristic evaluations of refraction coefficient (n), absorption coefficient (α), and electrical conductivity of epoxy resin in composite materials, which are present in fibers, were conducted by directly applying T-rays to FRP-laminated plates. Furthermore, T-ray scan imaging techniques were studied to detect defects in carbon fiber-reinforced plastics (CFRP) composites. Carbon fibers are conductive, whereas epoxy matrices are not conductive. Therefore, the interaction between the conductivity of carbon fibers in CFRP-laminated plates and T-ray applications were identified. Measurements regarding nanoparticle detection and sizing were also made by using microcavity cost and challenge based on spectra estimations [4,10–14].

Composite panels of honeycomb sandwiches, which are excellent in terms of their lightness, were used in this study. Foreign materials were inserted during composite manufacture, leading to decreased strength, shape change, and poor adhesion. After the carbon fibers were utilized as a skin, the foreign materials became not more visible than if a transparent glass fiber skin was used. Detecting these foreign materials is an important step toward checking the soundness of composite materials. This approach focused on a technique to monitor foreign materials on the honeycomb cell cores of these panels. Assuming that the foreign materials (e.g., foils) were present between the upper side of the honeycomb cells and a thin layer of carbon/glass fibers used as a skin while manufacturing the honeycomb sandwich panels, several artificial honeycomb sandwich panels with foreign materials were produced. Two kinds of skins were utilized and defined as "carbon skin" or "glass skin", with the skins being unidirectional for both carbon and glass fibers.

Experimental results of T-rays are herein presented for honeycomb sandwich composite panels. We also demonstrate a technique for measuring the refractive index (n), which is one of the properties of various materials that use THz waves, and consider a correlation between the angle of the electric field and the fiber angle of CFRP and GFRP composite panels to consider conductivity. The investigation regarding T-rays was found to successfully monitor the soundness of honeycomb sandwich composite panels. Two kinds of glass- and carbon-skin honeycomb sandwich composite panels were tested to examine honeycomb sandwich composite panel defects and optimal scanning methods were confirmed regarding the angle of the electric field versus the angle of fibers in glass and carbon skins for honeycomb sandwich composite panels.

2. Basic Theory Approach

Refractive Index Measurements

A reflection mode for T-rays is applied in the time domain to calculate a refractive index by analyzing the signals of T-rays reflected from the specimen, and a refractive index is induced by catching two surface and bottom signals reflected through the specimen. Figure 1 shows the direction of the T-ray. When a T-ray is reflected from the THz pulse emitter, a refractive index can be solved by utilizing time-of-flight (TOF) for a sample [1].

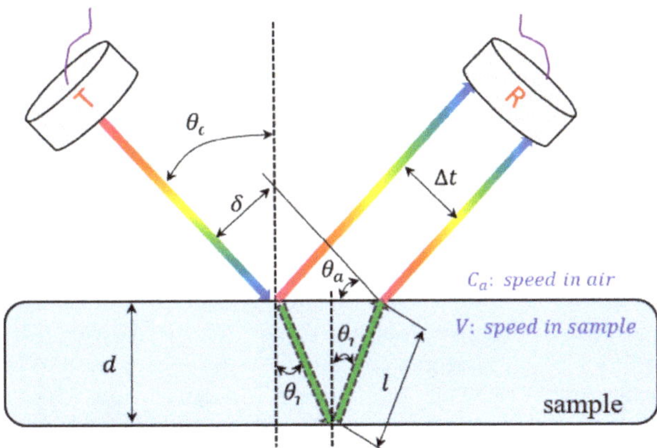

Figure 1. Schematic setup of the direction of a T-ray under reflection mode.

This reflection mode aims to acquire a refractive index by obtaining a length where the reflected fiber optics are passed through the upper and lower sides of the specimen out of the T-ray TOF.

Figure 1 shows the shape and direction of a T-ray. T in the figure refers to the transmitter, and R refers to the receiver; d means the thickness of the sample. Assuming that the T-ray is perpendicular to the sample, the time difference (Δt) can be calculated as follows [2–6]:

$$\Delta t = \frac{2d}{v} \tag{1}$$

As shown in Figure 1, the time difference (Δt) between the surface reflection wave and back reflector of the sample can be calculated by considering the time delay due to the sample thickness and the T-ray propagation path in reflection mode.

$$\Delta t = \frac{2l}{v} = \frac{\delta}{C_a} \tag{2}$$

where $l = d/cos\theta_r$, $\delta = 2lsin^2\theta_a = 2(d/cos\theta_r)sin\theta^2$, d refers to the specimen thickness, C_a refers to the T-ray speed in the air, n refers to the refractive index, and v refers to the T-ray speed inside the sample [2].

The resonance frequency (Δf) can be expressed as follows when the time delay due to the sample thickness and inclined T-ray path are tracked [2,14].

$$\Delta f = \frac{1}{\left(\frac{2d}{vcos\theta_r} - \frac{\delta}{C_a}\right)} = \frac{1}{\left(\frac{2d}{vcos\theta_r} - \frac{2dsin^2\theta_a}{cos\theta_a C_a}\right)} = \frac{1}{\frac{2d}{cos\theta_r}\left(\frac{1}{v} - \frac{sin^2\theta_a}{C_a}\right)} \tag{3}$$

where d refers to the sample thickness, θ_r refers to the inclination angle inside the sample, and θ_a refers to the inclination angle in the air. As presented above, the refractive index, which is one of the electromagnetic properties, can be obtained [2].

$$n^4 - An^2 - Asin^2\theta_{p1} = 0 \tag{4}$$

where A is $T^2V_{air}^2/4d_2^2$. In addition, T refers to the transmission time through the specimen and θ_{p1} denotes the inclination angle inside the sample.

The refractive index can be solved using Equation (5) under the transmission method [2].

$$\mathrm{n} = 1 + \frac{\Delta_t V_{air}}{t} \tag{5}$$

where Δ_t refers to the time delay between trials with and without sample and V_{air} refers to light speed in the air (3×10^{10}cm/s).

3. Experimental Device and Measurements

3.1. Measurement Device

Figure 2 shows the test method of nondestructive evaluation (NDE) THz-TDS to test sample characteristics. The NDE system collected and analyzed specimen characteristics and signals. The T-ray system in this experiment was made by Tera View (England). The NDE device consisted of a time–domain spectroscope (TDS) pulse tool and a frequency–domain continuous wave (CW) tool, which are TDS technologies to generate and control T-ray pulses and detect defects. The THz-TDS system acquired signals and valid data with a structural characteristic and optical device to adjust and control T-rays. The TDS device had a frequency bandwidth from 50 GHz to 4 THz, a window range of 300 ps, and was able to implement focal lengths for both 50 mm and 150 mm (full width at half maximum (FWHM)) of T-ray beams. In addition, the TDS system measured either under transmission or reflection mode (a pitch–catch technique with a small angle). The CW system had a frequency bandwidth of 50 GHz to 1.5 THz, with a focal length ranging from 50 mm to 150 mm. The TDS device was connected to the CW device via optical fibers. Figure 3 shows a simplified diagram of the T-ray system, where θ means a different angle surface fiber and E-field direction for the THz-TDS system.

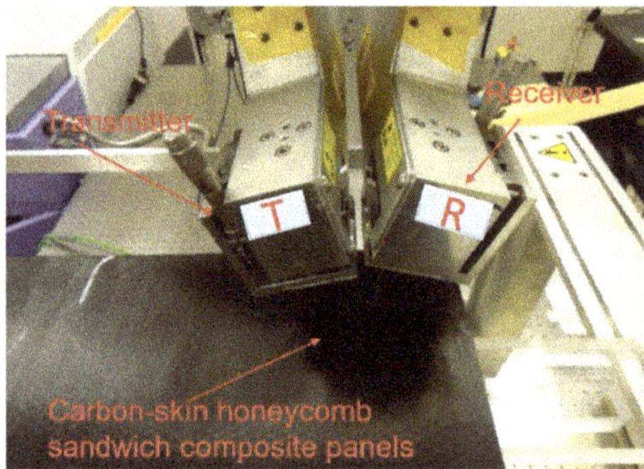

Figure 2. A photo of a T-ray system for measuring and imaging material properties under reflection mode.

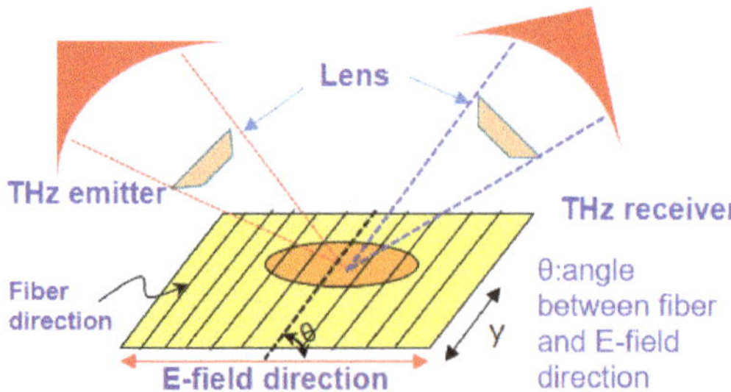

Figure 3. A simplified overview of the terahertz (THz) measurement method.

3.2. Measurement Technique

Figure 3 shows the T-ray measurement configuration under reflection mode. At the emitter, a T-ray was generated and sent to the receiver. A preferred sample was prepared with matching focal points for both the emitter and the receiver to conduct the experiment. The inclination angle of the T-ray lens was set to 16.6°. Figure 4 shows samples of carbon-skin and glass-skin honeycomb sandwich composite panels with foreign materials. For the case shown in Figure 4a, the foreign materials were invisible; however, the foreign materials in Figure 4b were visible. Figure 4c shows the locations of inserted foreign materials (brass foils). The dimensions of the foreign matters were 12.5 mm × 12.5 mm × 0.025 mm and 6.3 mm × 6.3 mm × 0.025 mm. Here, unidirectional carbon fibers were utilized and the diameter of fibers is 7 µm. The content of epoxy is around 37.0%.

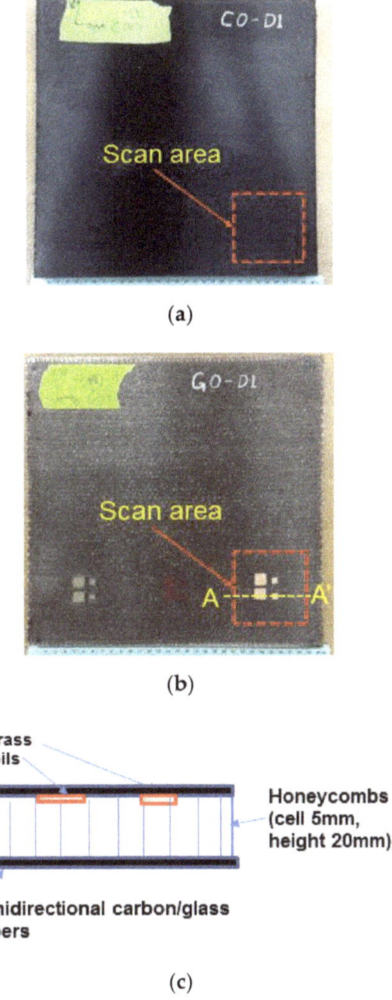

Figure 4. Samples of carbon-skin and glass-skin honeycomb sandwich composite panels with foreign materials. (**a**) Carbon-skin honeycomb composite panel with foreign materials. (**b**) Glass-skin honeycomb composite panel with foreign materials. (**c**) Stacking sequence of honeycomb composite panel at the A-A′ line.

4. Results and Discussion

4.1. Measurement of the Refractive Index

In order to measure the properties of the materials as T-ray parameters, T-ray pulses were acquired under transmission mode of GFRP composite materials. A time difference in reflection mode between the surface and bottom of the sample can clearly be seen in Figure 5. The GFRP specimen had a thickness of approximately 5.79 mm.

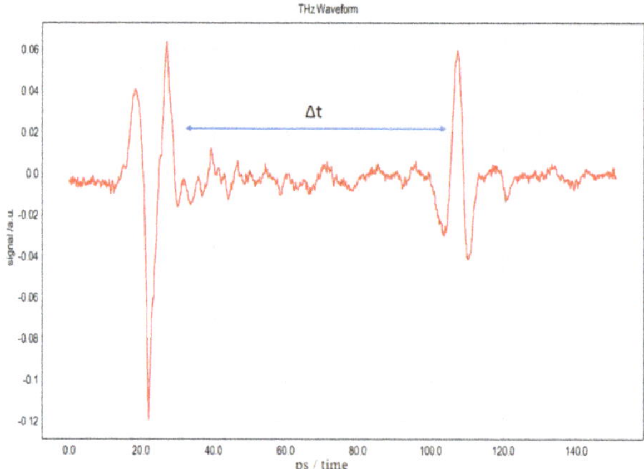

Figure 5. Terahertz time–domain spectroscopy (THz-TDS) pulses from a transmitted glass fiber-reinforced plastic (GFRP) sample (n = 1.95, Δt = 80.8 ps, thickness = 5.79 mm).

The time difference (Δt) measured using the reflective wave of the specimen was 80.8 ps. Thus, an optical time difference was easily obtained under reflection mode, which was one of the measurement techniques used to calculate the refractive index using Equation (2). The GFRP composite material and Poly methyl methacrylate (PMMA) and GFRP specimens were measured under reflection mode, as presented in Table 1. When compared to data in previous literature, less than 2.0% difference was revealed [6,12–15] due to the reflection mode measurement technique of T-rays during the experiment having unidirectional access considering a number of factors; therefore, the experiment was conducted conveniently. It was difficult to compare specimens with existing data because the GFRP samples were different from the existing samples in terms of manufacturing method and characteristics.

Table 1. Average T-ray refractive indices of the various materials.

Materials	Refractive Index (n) *	Refractive Index (n)
		Reflection Mode
Poly methyl methacrylate (PMMA)	1.60 ± 0.08	1.58 ± 0.07
Fused quartz	1.95 ± 0.05	1.95 ± 0.05
GFRPs	-	2.17 ± 0.05

Note: * Data in Refs. [3,9,10].

4.2. Evaluation on Electric Field(E-Fields) of Composite Materials

T-rays have limited transmitted power in conductive materials in contrast with nonconductive materials. Even if T-rays are applied and utilized in inspection of carbon fiber composites, in-depth studies are limited. In particular, carbon fiber-reinforced plastics (CFRP) consist of conductive carbon fibers and a nonconductive matrix. When the cross-section of CFRP composites is observed through a microscope, it can be seen to consist of various components, such as fibers and matrices, which significantly affect the conductivity. Because of this, a quantitative evaluation should be performed on characteristics of carbon fiber composites. According to previous study results, the electrical conductivity in the axial direction of a carbon fiber is approximately three orders of amount larger than in the radial direction of carbon fibers.

The CFRP composite is oriented in unidirectionally and the conductivity in a CFRP-laminated plate with various laminations is therefore affected. In particular, the mechanism that generates conductivity in the lateral direction (perpendicular to the fiber axis) depends on contact with fibers occurring between adjacent fibers. Few studies exist regarding the amount of electrical conductivity when using carbon fiber composites. Many studies [6,12–15] reported that the amount of conductivity (σ_l) in the axial direction was between 1 × 10^4 S/m and 6 × 10^4 S/m and the amount of conductivity (σ_t) in the radial direction was much wider, in the range of 2 S/m up to 600 S/m [16–18], as evidenced by Equation (6).

$$\sigma = \sigma_l cos^2\theta + \sigma_t sin^2\theta \quad (6)$$

Since the T-rays are much larger than the conductivity in the radial fiber direction ($\sigma_l \gg \sigma_t$), they can be significantly different according to the relative angle between the angle of carbon fibers and the angle of the electric field when transmitting through unidirectional CFRP composites [17,18]. When an electric T-ray field is parallel with the axial direction of the carbon fibers, conductivity becomes the largest value due to lower transmitted power to carbon fiber composites. However, when the direction of the electric field is perpendicular to the axis of carbon fibers, the conductivity decreases, whereas transmitted power becomes much greater. Using = 10^4 S/m, the skin depth of the unidirectional CFRP composite using T-rays could be 0.2 mm at 1 THz and 0.5 mm at 0.1 THz when the direction of the electric field is perpendicular to the axis of carbon fibers. The effect of transmitted power on angles of a 12-ply unidirectional CFRP composite-laminated plate was experimentally evaluated using a CW THz device. The T-ray transmitted power at the lower bandwidth (f—0.1 THz) of frequency was found to be greater than that of a noise level over 30 dB. Figure 6 shows the angular dependence of 0.1 THz reflection power. Figure 6a shows the angle between the carbon fiber orientation when the T-ray E-field was 0° and the peak-to-peak amplitude was approximately 2.25. Figure 6b shows the angle between the carbon fiber orientation when the T-ray E-field was 45.0° and the peak-to-peak amplitude was approximately 1.76. Figure 6c shows the angle between the carbon fiber orientation when the T-ray E-field was 90.0° and the peak-to-peak amplitude was approximately 1.70. The above results indicate that transmission power is easily achieved when the vector value in the T-ray E-field direction is 90.0°, thereby demonstrating lower peak-to-peak amplitude. On the other hand, when the vector value is 0°, the transmission power is difficult to generating, resulting in larger peak-to-peak amplitude of the reflection wave was larger and easier penetration.

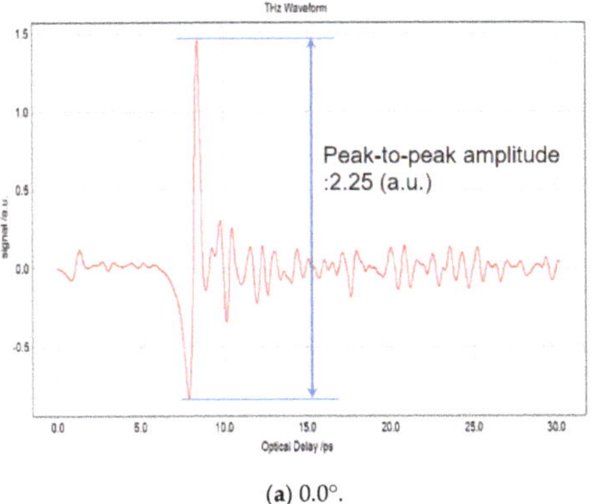

(**a**) 0.0°.

Figure 6. *Cont.*

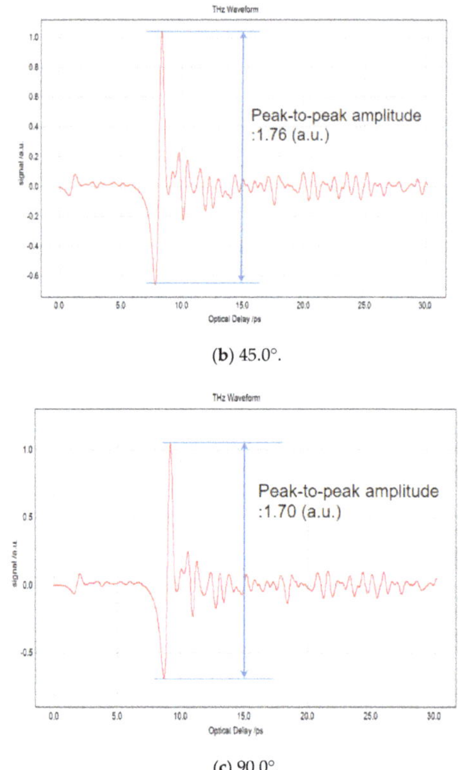

(**b**) 45.0°.

(**c**) 90.0°.

Figure 6. Angular dependence of reflection power of T-rays for carbon-skin honeycomb composite panels.

4.3. THz Imaging of Foreign Marerials in Honeycomb Sandwich Composite Panels

The brass foils attached to the bottom of the honeycomb sandwich composite panel using a carbon skin were detected to evaluate the characteristics of T-ray conductivity in the CFRP composite. The brass foils used here were 0.025 mm in thickness and two other dimensions of 12.5 mm × 12.5 mm and 6.3 mm × 6.3 mm. In particular, assuming that the direction of an electric field is perpendicular to the axial direction of carbon fibers and also parallel with the radial direction of carbon fibers, a different angle, θ, exists between the direction of the electric field and the direction of the carbon fibers. Defects were detected in this work under reflection mode using the TDS-THz system.

A correlation between the amount of conductivity and the S/N ratio of defective THz images can be expressed as $\sigma_l \gg \sigma_t$, with Equation (6) expressed as $\sigma \cong \sigma_l cos^2\theta$. The direction of the E-field and the direction of the carbon fibers were used to analyze the T-ray scan images, showing 1-ply carbon fiber defects in the prepreg sheet. The transmitted power of the T-rays penetrating the 1-ply sheet depends on an angle made between the fiber direction and E-field, allowing the T-ray to be transmitted and reflected from the surface of the sample according to the orientation of the fibers. When considering the simple resistance R equation, a single resistance body was made [6,17,18], allowing us to calculate the conductivity alongside the correlation of conductivity with the S/N ratio in defects from T-ray detection images.

The conductivity is at a minimum when the angle of an electric field is located at a 90° angle to the direction of carbon fibers in single-ply according to Equation $\sigma \cong \sigma_l cos^2\theta$. A signal-to-noise (S/N) ratio of defect image is at its greatest when this sample is positioned in this 90° angle. The S/N ratio could be at its worst at $0^0 (\sigma = 1.0\sigma_t)$ angle, which exhibited the largest conductivity. This defect detection method was determined to be valid between the S/N ratio based on a simple model for conductivity according to qualitative matching with the experimental image results.

Therefore, the defect detection resolution of T-rays could be different, depending on the relationship between the direction of the electric field and the direction of the carbon fibers. Figure 7 shows the scanning methods under reflection mode according to various angles made between the T-ray E-field and the fiber direction. Figure 7a shows the top view containing foreign materials. This figure shows the position of the four brass foils attached to the cell in the honeycomb sandwich composite panel using the T-ray scan method. Figure 7b shows the side view of the honeycomb sandwich composite panel, in which the position of the foreign materials between the cell of the honeycomb sandwich panel and the fiber skin can be seen. The dimensions of the foreign materials were 12.5 mm × 12.5 mm × 0.025 mm and 6.3 mm × 6.3 mm × 0.025 mm.

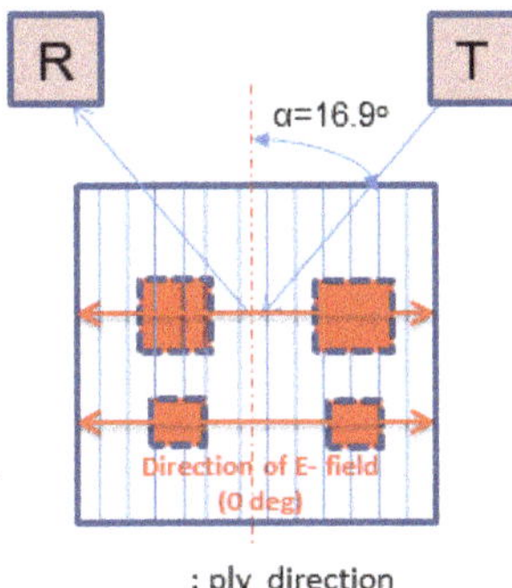

(**a**) Top view of honeycomb sandwiches.

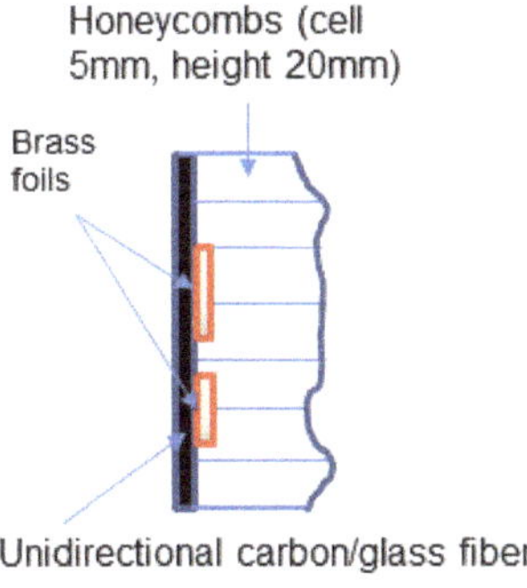

(**b**) Side view of honeycomb sandwiches.

Figure 7. Overview of T-ray reflection mode on honeycomb sandwich panels.

Figure 8 shows the implemented scan image using T-ray reflection mode. The specimen was the honeycomb sandwich composite panel with a glass fiber skin containing foreign materials processed between the glass fiber skin and the cell panel. The glass fiber orientation was changed to 0.0°, 45.0°, and 90.0° depending on the direction of the E-field to check the defect detection performance during scanning conduction. Figure 8a shows the scan image with an angle of 0.0° between the E-field and the glass fiber orientation, where four foreign defects were detected. Figure 8b,c show the scan images at 45.0° and 90.0°, where four foreign defects were detected. Thus, T-rays were transmitted through the glass fiber regardless of the T-ray direction due to the fiber glass used as a skin in the honeycomb sandwich panel being a nonconductive material and therefore not affecting the T-ray.

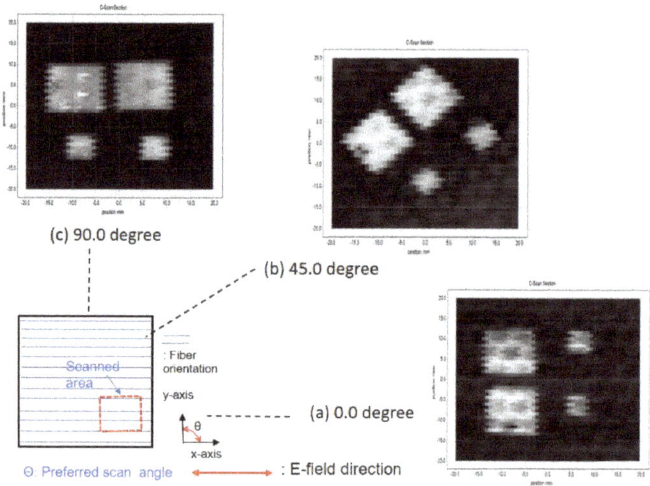

Figure 8. Terahertz scan images of glass-skin honeycomb sandwich composite panels with boned and brass foils (12.5 × 12.5 mm and 6.3 × 6.3 mm) at the bottom under time—domain spectroscopy (TDS) reflection mode.

Figure 9 shows the peak-to-peak amplitude-based scan image of a carbon-skin honeycomb sandwich composite panel by setting a time limit of the defection signals using T-ray reflection mode, as shown in Figure 7. The specimen was the honeycomb sandwich composite panel using a carbon fiber skin, in which the foreign matters were processed between the carbon fiber skin and the cell panel. The carbon fiber orientation was changed to 0.0°, 45.0° and 90.0° depending on the E-field direction, and scanning was conducted to check the defect detection performance, similar to that seen in Figure 8. Figure 9a shows the scan image at 0.0° between the E-field and glass fiber orientation, where four foreign defects could not be detected. The surface image only showed the honeycomb cell shape at a regular distance. Figure 9b shows the scan image at 45.0° between the E-field and the glass fiber orientation, where four foreign defects were detected, despite the S/N ratio being somewhat low. Figure 9c shows the scan image at 90.0° between the E-field and the carbon fiber orientation where four foreign defects were detected due to the presence of the highest S/N ratio.

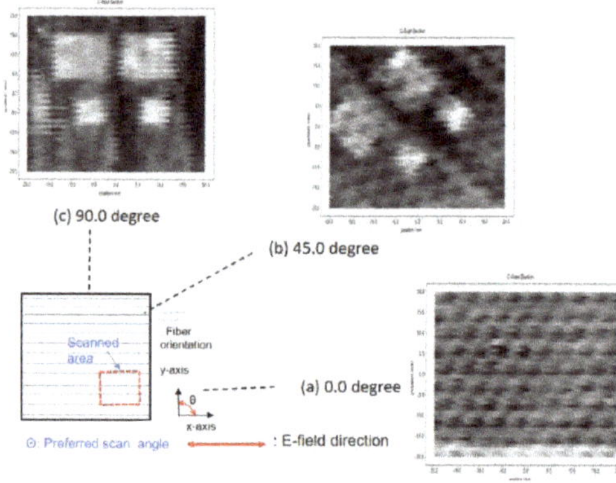

Figure 9. Terahertz scan images of carbon-skin honeycomb sandwiches composite panels with boned and brass foils (12.5 × 12.5 mm and 6.3 × 6.3 mm) at the bottom under TDS reflection mode.

The above results verify that foreign defect inspection of the honeycomb sandwich composite panel using a carbon fiber skin depends significantly on unidirectional carbon fiber locations according to the E-field direction. Figure 9 shows the TDS reflection mode scan image of defects which were observed in the carbon-skin honeycomb sandwich composite panel. Table 2 presents the conductivity levels of the carbon-skin honeycomb sandwich composite panel using the simplified model. Here, θ means the angle between the direction of an axial one-ply fiber and the direction of the electric field and ψ means the angle between the direction of an axial second-ply fiber and the direction of the electric field. These angles used were (a) θ = 0.0°, (b) θ = 45.0°, and (c) θ = 90.0°. As shown in Equation (6), the resistance became minimal when the angle between the E-field and the carbon fiber orientation was 90.0°. In particular, the resistance became maximal when the angle was 0°. Thus, the T-ray could not penetrate the carbon-skin honeycomb sandwich composite panel at all, so the defects could not be detected, as shown in Figure 9a.

Table 2. Predicted simple resistor model of one-ply conductivity.

Resistance	Angles		
	0°	**45.0°**	**90.0°**
θ	0°	45.0°	90.0°
σ1	1	0.5	0
σ	1.0 $\cos^2\theta$	0.5 $\cos^2\theta$	0 $\cos^2\theta$
Req	1	0.25	0

Figure 10 shows the resistance size of the E-field reception according to the carbon fiber orientation of the carbon-skin honeycomb sandwich composite panel. In particular, since the resistance was smallest when the angle between the carbon fiber orientation of the carbon-skin honeycomb sandwich composite panel and the E-field was 90.0°, the defect signals could be optimized because the transmission rate of the T-ray was very high.

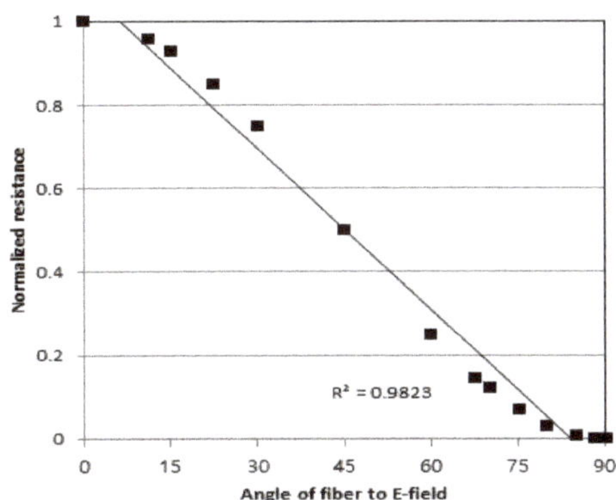

Figure 10. Relationship between the angle of the fiber and the E-field and normalized resistance in a unidirectional carbon composite laminate.

Therefore, when the CFRP skins were utilized as the top layers for the honeycomb sandwich composite panels, the foreign materials in the scanning images were most visible at 90.0°; however, the foreign materials could not be clearly observed at other angles, particularly 0.0°, because the T-rays were unable to penetrate the CFRP skin when the top surface fibers of the honeycomb sandwich composite panels were stacked.

5. Conclusions

This study aimed to investigate the application and utilization of a technique to measure the refractive index (n), which is one of the material properties in various materials using THz waves, and consider a correlation between the angle of an electric field and the fiber angles of CFRP and GFRP composite panels to analyze conductivity. T-ray investigations were successfully performed to monitor the soundness of honeycomb sandwiches composite panels, allowing us to obtain the following THz results of the honeycomb sandwich composite panels:

(1) The refractive index can be measured under reflection mode using T-rays, particularly with glass-fiber composite materials.
(2) T-rays have limited transmission power to some degree for carbon-skin honeycomb sandwich composite panels, but T-ray transmission power was found to be related to the vector angles between the direction of the E-field and the direction of carbon fibers, thereby affecting the electrical conductivity of the composite panels and the E-field direction.
(3) For the glass-skin honeycomb sandwich composite panels, scan images can be obtained using T-rays regardless of the E-field direction due to nonconducting materials. When utilizing carbon skins, optimal experimental techniques based on peak-to-peak amplitude can be acquired approximately at 90° between the direction of the E-field and the direction of carbon fibers, because T-rays can easily penetrate carbon fiber skins in honeycomb sandwich composite panels.
(4) If T-ray systems can be produced to be inexpensive and portable, they may be highly useful and potentially nondestructive inspection tools in future applications.

Author Contributions: K.-H.I. suggested and designed the experiments; S.-K.K., Y.-T.C. and Y.-D.W. performed the experiments; C.-P.C. and J.-A.J. helped in the accomplishment of ideas and the administration of the experiments. The data were discussed and analyzed and the manuscript was written and revised by all members. All authors have read and agreed to the published version of the manuscript.

Funding: This research was supported by Basic Science Research Program through the National Research Foundation of Korea (NRF) (No. 2018R1D1A1B07049775) and also experimentally helped by the CNDE at Iowa State University, USA.

Conflicts of Interest: The authors declare no conflict of interest.

References

1. Chiou, C.P.; Blackshire, J.L.; Thompson, R.B.; Hu, B.B. Terahertz Ray System Calibration and Material Characterzations. *Rev. QNDE* **2009**, *28*, 410–417.
2. Im, K.H.; Lee, K.S.; Yang, I.Y.; Yang, Y.J.; Seo, Y.H.; Hsu, D.K. Advanced T-ray Nondestructive Evaluation of Defects in FRP Solid Composites. *Inter. J. Precis. Eng. Manuf.* **2013**, *14*, 1093–1098. [CrossRef]
3. Li, M.W.; Liang, C.P.; Zhang, Y.B.; Yi, Z.; Chen, X.F.; Zhou, Z.G.; Yang, H.; Tang, Y.J.; Yi, Y.G. Terahertz Wideband Perfect Absorber Based on Open Loop with Cross Nested Structure. *Results Phys.* **2019**, *15*, 102603. [CrossRef]
4. Chunlian, C.; Zhang, Y.; Chen, X.; Yang, H.; Yi, Z.; Yao, W.; Tang, Y.; Yi, Y.; Wang, J.; Wu, P. A Dual-Band Metamaterial Absorber for Graphene Surface Plasmon Resonance at Terahertz Frequency. *Phys. E Low-Dimens. Syst. Nanostruct.* **2020**, *117*, 113840. [CrossRef]
5. Zhang, Y.; Wu, P.; Zhou, Z.; Chen, X.; Yi, Z.; Zhu, J.; Zhang, T.; Jile, H. Study on Temperature Adjustable Terahertz Metamaterial Absorber Based on Vanadium Dioxide. *IEEE Access* **2020**, *8*, 85154–85161. [CrossRef]
6. Im, K.-H.; Kim, S.-K.; Jung, J.-A.; Cho, Y.-T.; Woo, Y.-D.; Chiou, C.-P. NDE Detection Techniques and Characterization of Aluminum Wires Embedded in Honeycomb Sandwich Composite Panels Using Terahertz Waves. *Materials* **2019**, *12*, 1264. [CrossRef] [PubMed]
7. Hsu, D.K.; Lee, K.S.; Park, J.W.; Woo, Y.D.; Im, K.H. NDE Inspection of Terahertz Waves in Wind Turbine Composites. *Inter. J. Precis. Eng. Manuf.* **2012**, *13*, 1183–1189. [CrossRef]
8. Huber, R.; Brodschelm, A.; Tauser, A.; Leitenstorfer, A. Generation and Field-Resolved Detection of Femtosecond Electromagnetic Pulses Tunable up to 41 THz. *Appl. Phys. Lett.* **2000**, *76*, 3191–3199. [CrossRef]

9. Rudd, J.V.; Mittleman, D.M. Influence of Substrate-Lens Design in Terahertz Time-Domain Spectroscopy. *J. Opt. Soc. Am. B* **2000**, *19*, 319–329. [CrossRef]
10. Gregory, I.S.; Baker, C.; Tribe, W.; Bradley, I.V.; Evans, M.J.; Linfield, E.H. Optimization of Photomixers and Antennas for Continuous-Wave Terahertz Emission. *IEEE J. Quantum Electron.* **2000**, *41*, 717–728. [CrossRef]
11. Brown, E.R.; Smith, F.W.; McIntosh, K.A. Coherent Millimeterwave Generation by Heterodyne Conversion in Low-Temperature-Grown GaAs Photoconductors. *J. Appl. Phys.* **1993**, *73*, 1480–1484. [CrossRef]
12. Brown, E.R.; McIntosh, K.A.; Nichols, K.B.; Dennis, C.L. Photomixing up to 3.8 THz in Low-Temperature-Grown GaAs. *Appl. Phys. Lett.* **1995**, *66*, 285–287. [CrossRef]
13. Zhang, J.; Li, W.; Cui, H.L.; Shi, C.; Han, X.; Ma, Y.; Chen, J.; Chang, T.; Wei, D.; Zhang, Y.; et al. Nondestructive Evaluation of Carbon Fiber Reinforced Polymer Composites Using Reflective Terahertz Images. *Sensors* **2016**, *16*, 875. [CrossRef]
14. Li, L.; Yang, W.; Shui, T.; Wang, X. Ultrasensitive Sizing Sensor for a Single Nanoparticle in a Hybrid Nonlinear Microcavity. *IEEE Photonics J.* **2020**, *12*, 1–8. [CrossRef]
15. Wu, P.; Chen, Z.; Jile, H.; Zhang, C.; Xu, D.; Lv, L. An Infrared Perfect Absorber Based on Metal-Dielectric-Metal Multi-Layer Films with Nanocircle Holes Arrays. *Results Phys.* **2020**, *16*, 102952. [CrossRef]
16. Schueler, R.; Joshi, S.P.; Schulte, K. Damage Detection in CFRP by Electrical Conductivity Mapping. *Compos. Sci. Technol.* **2001**, *61*, 921–930. [CrossRef]
17. Hsu, D.K. Characterization of a Graphite/Epoxy Laminate by Electrical Resistivity Measurements. *Rev. QNDE* **1985**, *4*, 1219–1228.
18. Tse, K.W.; Moyer, C.A.; Arajs, S. Electrical Conductivity of Graphite Fiber epoxy Resin Composites. *Mater. Sci. Eng.* **1981**, *49*, 41–46. [CrossRef]

Article

A 350-GHz Coupled Stack Oscillator with −0.8 dBm Output Power in 65-nm Bulk CMOS Process

Thanh Dat Nguyen and Jong-Phil Hong *

School of Electrical Engineering, Chungbuk National University, Cheongju 28644, Korea; ntdat@cbnu.ac.kr
* Correspondence: jphong@cbnu.ac.kr

Received: 15 July 2020; Accepted: 27 July 2020; Published: 28 July 2020

Abstract: This paper presents a push-push coupled stack oscillator that achieves a high output power level at terahertz (THz) wave frequency. The proposed stack oscillator core adopts a frequency selective negative resistance topology to improve negative transconductance at the fundamental frequency and a transformer connected between gate and drain terminals of cross pair transistors to minimize the power loss at the second harmonic frequency. Next, the phases and the oscillation frequencies between the oscillator cores are locked by employing an inductor of frequency selective negative resistance topology. The proposed topology was implemented in a 65-nm bulk CMOS technology. The highest measured output power is −0.8 dBm at 353.2 GHz while dissipating 205 mW from a 2.8 V supply voltage.

Keywords: oscillator; THz; high output power; CMOS

1. Introduction

The terahertz (THz) frequency range, which is from 300 GHz to 3 THz, has recently gained much attention from researchers due to its wide range of applications such as high-speed communication, imaging security system, and spectroscopy [1,2]. In these applications, a high power and high frequency signal source is one of the most important components to create a system with superior quality. Among the technologies for designing a signal source, CMOS technology stands out as a reliable selection to build a high-quality signal source because of its low production cost and compact size. However, CMOS signal sources present some difficulties such as low maximum oscillation frequency (f_{max}) and low output power at the THz frequency range.

Though a fundamental frequency oscillator is widely adopted for generating an output signal having oscillation frequency smaller than f_{max} [3–7], but a fundamental frequency oscillator cannot generate a high frequency output signal that has oscillation frequency greater than f_{max}. Therefore, a harmonic frequency oscillator is a viable solution to overcome the low f_{max} of CMOS technology and to generate a high frequency output signal that has oscillation frequency greater than f_{max}. Nevertheless, only a limited output power extracted from a single oscillator core is a problem of harmonic frequency oscillator at this high frequency range [8,9].

This paper proposes a THz frequency CMOS coupled stack oscillator with a good output power level. The oscillator core uses a frequency selective negative resistance (FSNR) tank to increase the negative transconductance and the transformer-based topology to minimize the loss of the second harmonic power. The push-push topology is employed to generate a high oscillation frequency over f_{max} and multiple coupled oscillator cores to obtain a high output power. The proposed oscillator, implemented in a 65 nm bulk CMOS process, generates a maximum output power of −0.8 dBm at 350 GHz. In Section 2, the proposed signal source's structure is analyzed. Section 3 presents the measurement setups and measurement results of the proposed oscillator. Finally, Section 4 summarize the findings in this paper.

2. The Proposed Signal Source

Figure 1a shows a schematic of a conventional cross coupled oscillator (XCO). At the THz frequency, a conventional XCO suffers from low negative transconductance at the fundamental frequency and high loss at the second harmonic frequency because of the direct connection of a low gate impedance of transistors M_1 and M_2 to the second harmonic output path. In Figure 1b, a conventional stack oscillator with FSNR tank, implemented by transistors M_3, M_4, and inductor L_2, is connected in parallel with the cross-coupled pair M_1 and M_2 to boost the total negative transconductance at the fundamental frequency, so a higher output power can be generated [2]. A novel stack oscillator, shown in Figure 1c, is proposed to minimize the loss at the second harmonic frequency of the conventional stack oscillator by connecting a transformer between gate and drain terminals of cross-coupled pair.

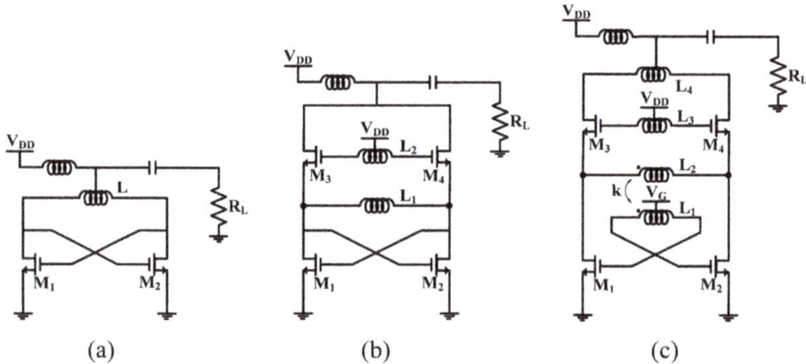

Figure 1. Schematics of (**a**) a conventional cross coupled oscillator (XCO), (**b**) a conventional stack oscillator, and (**c**) a proposed stack oscillator.

Figure 2 shows circuit simulation results for a negative transconductance at 180 GHz fundamental frequency of the conventional XCO, the conventional stack oscillator, and the proposed stack oscillator. All topologies have the same of transistor size W/L of 12 µm/60 nm, and the same voltage across the drain and source terminals of each transistor of 1V. The conventional XCO generates only 2 mS negative transconductance, and conventional stack oscillator generates 11.9 mS negative transconductance due to the extra negative transconductance added by FSNR tank. The negative transconductance of the proposed stack oscillator increases from 5.8 mS to 11.6 mS with a coupling factor k increases from 0.2 to 0.9.

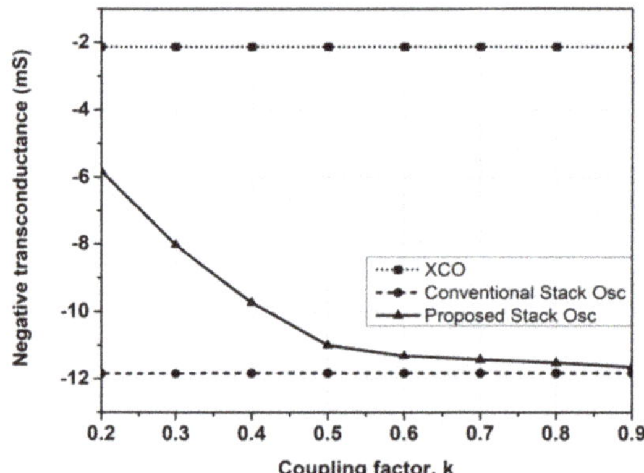

Figure 2. Circuit simulation results of negative transconductance at the fundamental frequency of conventional XCO, conventional stack oscillator, and proposed stack oscillator.

The effectiveness of the transformer in the proposed stack oscillator in reducing the loss of the second harmonic output power is simulated and shown in Figure 3. In this simulation, we assume that both the conventional stack oscillator and the proposed stack oscillator generate 0 dBm power at the second harmonic frequency inside the oscillator tank. The gate impedance looking from the drain terminals of transistors in the proposed stack oscillator is increased with a decrease of the coupling factor k and increased from 3.5 Ω at k = 0.9 up to 39.7 Ω at k = 0.2 compared with that of the conventional stack oscillator. Therefore, the second harmonic output power of the proposed stack oscillator is improved from 0.4 dB at k = 0.9 up to 4.6 dB at k = 0.2 compared with that of the conventional stack oscillator.

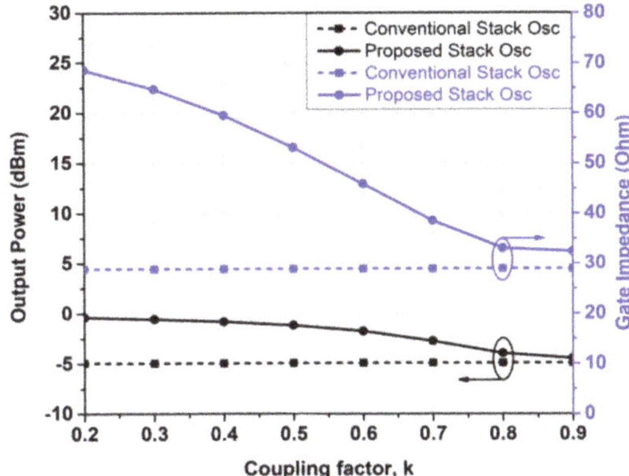

Figure 3. Circuit simulation results of output power and gate impedance of conventional stack oscillator and proposed stack oscillator at the second harmonic frequency.

A combination of the FSNR tank and the transformer connected between gate and drain terminals of the cross-coupled pair in the proposed stack oscillator boosts the second harmonic output power. Figure 4 shows that the second harmonic output power of the proposed stack oscillator is higher than that of the conventional stack oscillator when k > 0.5. Since the minimum space between two metals at a top metal layer of the design process is 2 μm, a simulation result from the high frequency structure simulator (HFSS) that is one of ANSYS products [10] shows that a coplanar transformer has a maximum coupling factor k of 0.65. At k = 0.65, the output power of the proposed stack oscillator is higher than the output power of the conventional stack oscillator 1.7 dB, and the output power of the conventional XCO 14.6 dB.

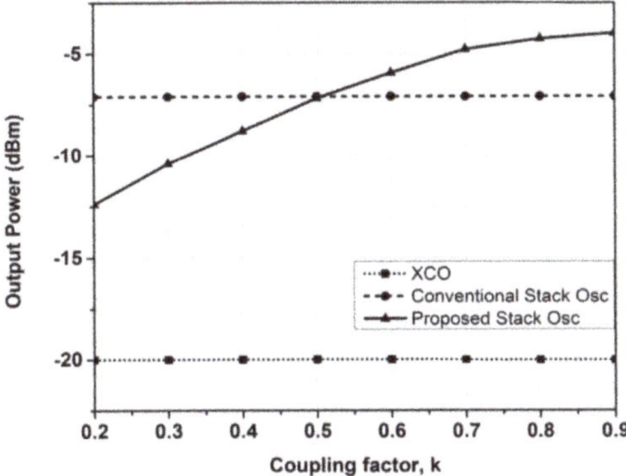

Figure 4. Circuit simulation results of the second harmonic output power of XCO, conventional stack oscillator, and proposed stack oscillator.

To further improve the second harmonic output power, we propose a coupled stack oscillator, shown in Figure 5, with oscillator cores are the presented stack oscillators. To lock phase differences and oscillation frequencies between oscillator cores, the inductor L_3 of the proposed stack oscillator is split into two inductors with the same inductance value of $L_3/2$. These inductors are implemented as transmission lines TL1. One transmission line TL1 is connected to a gate of a FSNR transistor of the previous oscillator core, and another transmission line TL1 is connected to a gate of a FSNR transistor of the next oscillator core. A resistor R_G is connected between two transmission lines TL1 to provide a gate bias voltage path for transistors M_3 and M_4, and to guarantee a differential operation between the adjacent oscillator cores. Transmission lines TL2 that are implemented at the drain terminals of transistors M_3 and M_4 allow a larger voltage swing at the fundamental frequency signal, so a higher output power can be obtained. Transmission lines TL3 combine the output power which is generated from each oscillator cores to an output port and perform output impedance matching at the second harmonic frequency.

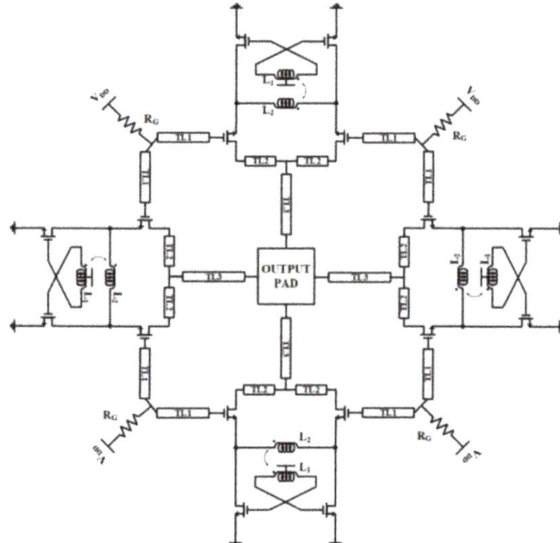

Figure 5. Schematic of the proposed coupled stack oscillator.

3. Measurement Results

The proposed oscillator was fabricated in a 65 nm bulk CMOS process. Figure 6 shows a chip micrograph of the proposed stack coupled oscillator with a total implementation area of 549×468 μm^2. The output spectrum of the proposed oscillator was measured based on the spectrum measurement setup using an R&S FSW26 signal and spectrum analyzer, as shown in Figure 7a. A GGB DC Probe was connected to the power pads of the fabricated chip. The GGB DC probe with a GPPG pin configuration provided two paths to supply two different DC voltage levels that are a supply voltage VDD and a gate bias voltage VG to operate the fabricated circuit. The output pads were connected to a GGB model 500 B probe with GSG pin configuration to extract the output power. After that, the output power was conducted to a Farran WR-2 harmonic mixer. The Farran WR-2 down-conversion is a harmonic mixer with interface WR2 has a function of down-converting the frequency of the input signal by mixing input signal with a local oscillator (LO) signal. At the harmonic mixer, the output signal from the proposed stacked coupled oscillator was down-converted by mixing it with a LO signal generated from an R&S FSW26 signal and spectrum analyzer. The down-converted output signal from the harmonic mixer was received and the oscillation frequency was automatically calculated by the R&S FSW26 signal and spectrum analyzer. Figure 8a shows the measured oscillation frequency of

350 GHz. Figure 8b shows an oscillation frequency range of the proposed oscillator with a change of the supply voltage VDD from 2 V to 2.8 V and the gate bias voltage VG from 1 V to 1.4 V. As shown in Figure 8b, the oscillation frequency of the proposed oscillator increases with an increase of the supply voltage VDD and decrease with an increase of the gate bias voltage VG. The proposed oscillator has a minimum oscillation frequency of 345 GHz at VDD = 2 V and VG = 1.4 V and a maximum oscillation frequency of 353.2 GHz at VDD = 2.8 V and VG = 1 V.

Figure 7b shows a power measurement setup of the proposed stack coupled oscillator. The output power of the proposed coupled stack oscillator was measured by a PM5 power meter. The output pad was connected to a GGB model 500 B probe, a waveguide bend WR3, a waveguide tapper WR3.4-WR10, and a waveguide WR10 with insertion losses of 4 dB, 1 dB, 0.4 dB, and 0.3 dB, respectively. As shown in Figure 8c, the highest measured output power of the proposed coupled stack oscillator is 0.832 mW or −0.8 dBm at a supply voltage VDD of 2.8 V and a gate bias voltage VG of 1 V while the proposed stack coupled oscillator consumes 205 mW of DC power. Figure 8d shows an output power range of the proposed coupled oscillator. As shown in Figure 8d, the output power of the proposed coupled oscillator increases with an increase of the supply voltage VDD and decreases with an increase of the gate bias voltage VG. From overall oscillation frequency and output power measurement results, the proposed stacked coupled oscillator obtains a high frequency of 353.2 GHz with a high output power of −0.8 dBm at the same condition that is a high supply voltage VDD of 2.8 V and a low gate bias voltage VG of 1 V.

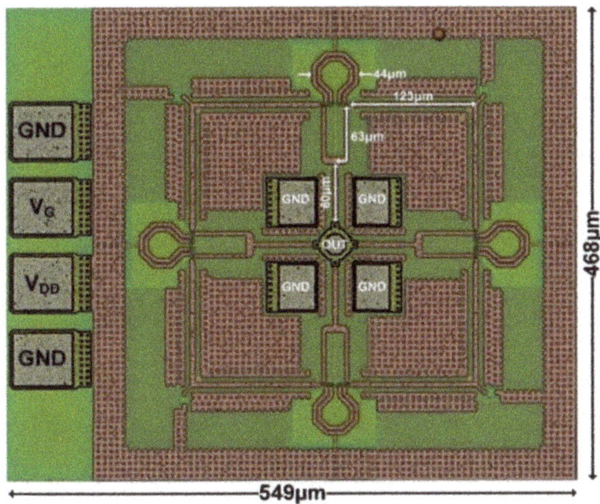

Figure 6. Chip micrograph of the proposed coupled stack oscillator.

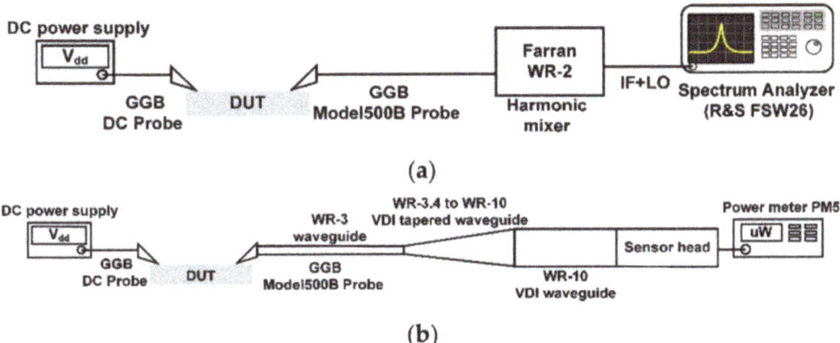

Figure 7. (**a**) Output spectrum measurement setup; (**b**) output power measurement setup.

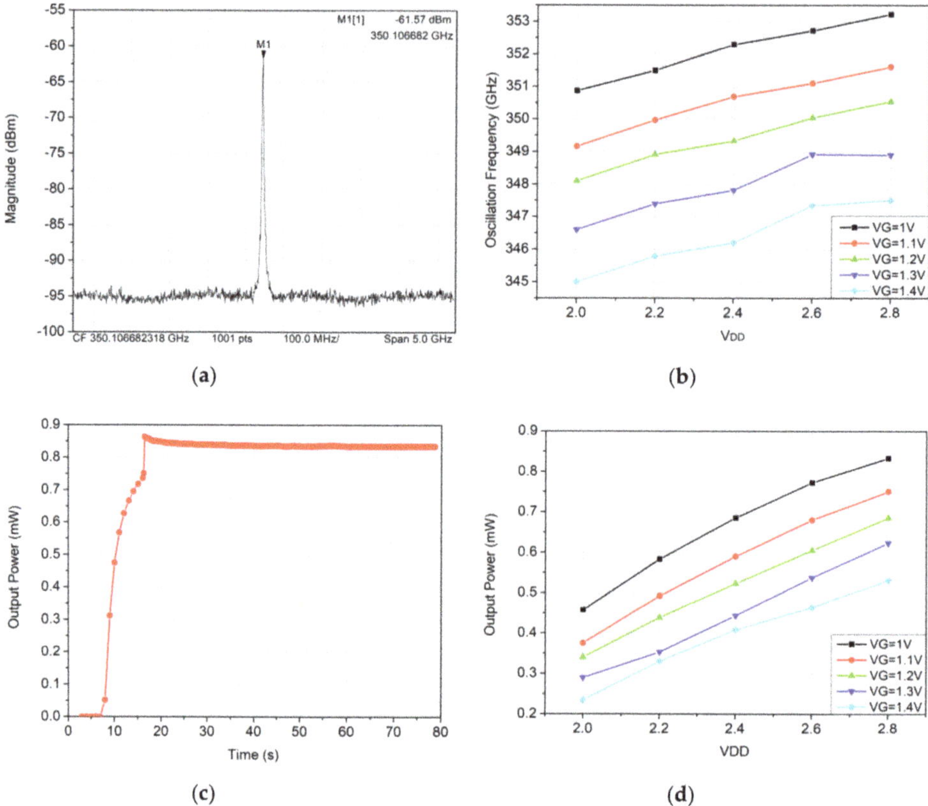

Figure 8. (**a**) Measured output spectrum, (**b**) oscillation frequency range, (**c**) measured the highest output power and (**d**) output power range of the proposed coupled stack oscillator.

Table 1 shows a comparison of the performance of the proposed oscillator with the state-of-the-art THz oscillators. Oscillators in [11–13] generate output signals at the fourth harmonic frequency, while the work in [14] generates an output signal at the second harmonic frequency with an unstack core oscillator structure. Similar to [14], this work also generates an output signal at the second harmonic frequency but with a stack core oscillator structure. Both a spectrum analyzer and a power meter can measure output power of the proposed oscillator, but the power measurement accuracy of each measurement device is different. Spectrum analyzer can accurately measure frequency spectrum but cannot accurately measure output power generated from oscillators. The reason of low accuracy of power measurement of a spectrum analyzer is that there is a harmonic mixer with a high loss of approximately 50 dBm in the spectrum measurement setup. This decreases power sensitivity of spectrum analyzer. In contrast, a power meter cannot measure frequency but can accurately measure output power generated from oscillators. The reason of high accuracy of power measurement of a power meter is that loss from measurement devices in the power measurement setup is low around several dBm. Therefore, power meter can show an accurate power measurement when calibration function is correctly applied. Because of these reasons, the output power of the proposed oscillator is −0.8 dBm. Among the state-of-the-art THz oscillators operating from 300 GHz to around 400 GHz, the proposed oscillator shows the highest output power of −0.8 dBm and the smallest implementation area of 0.25 mm^2 while it has a high DC-to-RF efficiency of 0.41% at a high oscillation frequency of 350 GHz.

Table 1. Performance comparison with state-of-the-art terahertz (THz) oscillators.

	[11]	[12]	[13]	[14]	This Work
Frequency (GHz)	380	338	312	417	353.2
Type	4th harmonic	4th harmonic	4th harmonic	2nd harmonic	2nd harmonic
P_{OUT} Max. (dBm)	−17.3	−0.9	0.8	−2.5	−0.8
No. of cores	1	16	4	4	4
P_{DC} (mW)	380	380	298	108	205
DC-to-RF eff. (%)	0.41	0.005	0.42	0.5	0.41
Chip Area (mm^2)	4.1	3.9	0.36	0.36	0.25
Technology	130 nm SiGe	65 nm CMOS	65 nm CMOS	65 nm CMOS	65 nm CMOS
No. of cores	1	16	4	4	4

4. Conclusions

This paper introduced a high-power THz CMOS signal source. The proposed signal source can generate a high output power level of −0.8 dBm at the THz frequency by employing an FSNR topology to increase the negative transconductance at the fundamental frequency and proposing a transformer which is connected between gate and drain terminals of cross-coupled pair to minimize the power loss at the second harmonic frequency. A coupled topology with the proposed coupled structure was also adopted to further increase the output power. Based on the verified measurement results, the proposed signal source is recommended operating with a supply voltage range from 2 V to 2.8 V.

Author Contributions: Conceptualization, T.D.N. and J.-P.H.; methodology, T.D.N. and J.-P.H.; circuit analysis, T.D.N. and J.-P.H.; investigation, T.D.N. and J.-P.H.; writing—original draft preparation, T.D.N.; writing—review and editing, T.D.N. and J.-P.H. All authors have read and agreed to the published version of the manuscript.

Funding: This work was supported by the Korea Institute for Advancement of Technology (KIAT) grant funded by the Korea Government (MOTIE) (N0001883, The Competency Development Program for Industry Specialist). This work was also supported by Basic Science Research Program through the National Research Foundation of Korea (NRF) funded by the Ministry of Education (NRF-2018R1D1A1B07042607).

Conflicts of Interest: The authors declare no conflict of interest. The funders had no role in the design of the study; in the collection, analyses, or interpretation of data; in the writing of the manuscript, or in the decision to publish the results.

References

1. Koo, H.; Kim, C.-Y.; Hong, S. Design and analysis of 239 GHz CMOS push-push transformer-based VCO with high efficiency and wide tuning range. *IEEE Trans. Circuits Syst. I Regul. Pap.* **2015**, *62*, 1883–1893. [CrossRef]
2. Gu, Q.; Xu, Z.; Jian, H.-Y.; Pan, B.; Xu, X.; Chang, M.-C.F.; Liu, W.; Fetterman, H. CMOS THz generator with frequency selective negative resistance tank. *IEEE Trans. Terahertz Sci. Technol.* **2012**, *2*, 193–202. [CrossRef]
3. Kwon, H.-T.; Nguyen, D.; Hong, J.-P. A 219-GHz fundamental oscillator with 0.5 mW peak output power and 2.08% DC-to-RF efficiency in a 65 nm CMOS. In Proceedings of the 2016 IEEE MTT-S International Microwave Symposium (IMS), San Francisco, CA, USA, 22–27 May 2016. [CrossRef]
4. Nguyen, T.D.; Park, H.; Hong, J.-P. A millimeter-wave fundamental frequency CMOS-based oscillator with high output power. *Electronics* **2019**, *8*, 1228. [CrossRef]
5. Li, C.-H.; Ko, C.-L.; Kuo, C.-N.; Kuo, M.-C.; Chang, D.-C. A 340 GHz triple-push oscillator with differential output in 40 nm CMOS. *IEEE Microw. Wirel. Compon. Lett.* **2014**, *24*, 863–865. [CrossRef]
6. Grzyb, J.; Zhao, Y.; Pfeiffer, U.R. A 288-GHz lens-integrated balanced triple-push source in a 65 nm CMOS technology. *IEEE J. Solid State Circuits* **2013**, *48*, 1751–1761. [CrossRef]
7. Samuel, J.; Eliezer, H.; Eran, S. A 300 GHz wirelessly locked 2 × 3 array radiating 5.4 dBm with 5.1% DC-to-RF efficiency in 65nm CMOS. In Proceedings of the 63rd IEEE International Solid-State Circuits Conference (ISSCC), San Francisco, CA, USA, 31 January–4 February 2016; pp. 348–349.
8. Kananizadeh, R.; Momeni, O. A 190-GHz VCO with 20.7% tuning range employing an active mode switching block in a 130 nm SiGe BiCMOS. *IEEE J. Solid State Circuits* **2017**, *52*, 1–11. [CrossRef]

9. Khamaisi, B.; Socher, E. A 159–169 GHz frequency source with 1.26 mW peak output power in 65 nm CMOS. In Proceedings of the European Microwave Integrated Circuits Conference, Nuremberg, Germany, 6–10 October 2013; pp. 536–539.
10. 3D Electromagnetic Field Simulator for RF and Wireless Design. Available online: https://www.ansys.com/products/electronics/ansys-hfss (accessed on 27 July 2020).
11. Park, J.-D.; Kang, S.; Niknejad, A.M. A 0.38 THz fully integrated transceiver utilizing a quadrature push-push harmonic circuitry in SiGe BiCMOS. *IEEE J. Solid State Circuits* **2012**, *47*, 2344–2354. [CrossRef]
12. Tousi, Y.; Afshari, E. A high-power and scalable 2-D phased array for terahertz CMOS integrated systems. *IEEE J. Solid State Circuits* **2014**, *50*, 597–609. [CrossRef]
13. Wu, L.; Liao, S.; Xue, Q. A 312 GHz CMOS injection locked radiator with chip and package distributed antenna. *IEEE J. Solid State Circuits* **2017**, *52*, 2920–2933. [CrossRef]
14. Nguyen, T.D.; Hong, J.-P. A −2.5-dBm, 5.1%-tuning-range, 417-GHz signal source with gate-to-drain-coupled oscillator in 65-nm CMOS process. *IEEE Microw. Wirel. Compon. Lett.* **2018**, *28*, 1023–1025. [CrossRef]

Article

Terahertz Displacement Sensing Based on Interface States of Hetero-Structures

Lan-Lan Xu [1,2], Ya-Xian Fan [1,3,*], Huan Liu [1], Tao Zhang [1] and Zhi-Yong Tao [1,3,*]

1. Guangxi Key Laboratory of Wireless Wideband Communication and Signal Processing, Guilin University of Electronic Technology, Guilin 541004, China; lanlanxu@hrbeu.edu.cn (L.-L.X.); huanliu@guet.edu.cn (H.L.); zt@guet.edu.cn (T.Z.)
2. Key Lab of In-Fiber Integrated Optics, Ministry Education of China, Harbin Engineering University, Harbin 150001, China
3. Academy of Marine Information Technology, Guilin University of Electronic Technology, Beihai 536000, China

* Correspondence: yxfan@guet.edu.cn (Y.-X.F.); zytao@guet.edu.cn (Z.-Y.T.)

Received: 10 July 2020; Accepted: 27 July 2020; Published: 28 July 2020

Abstract: Herein, we propose a nano displacement sensor based on the interface state of a terahertz hetero-structure waveguide. The waveguide consists of two periodically corrugated metallic tubes with different duty ratios, which can result in similar forbidden bands in their frequency spectra. It was found that the topological properties of these forbidden bands are different, and the hetero-structure can be formed by connecting these two waveguides. In the hetero-structure waveguide, the interface state of an extraordinary transmission can always arise within the former forbidden bands, the peak frequency of which is highly dependent on the cavity length at the interface of the two periodic waveguides. So, by carefully designing the structure's topological property, the hetero-structure waveguide can be efficiently used to produce a displacement sensor in the THz frequency range. The simulations show that the resolution of the displacement can be as small as 90 nm and the sensitivity can reach over 1.2 GHz/μm. Such a sensitive interface state of the proposed hetero-structure waveguide will greatly benefit THz applications of functional devices, including not only displacement sensors but also switches with high extinction ratios, tunable narrow-band filters, and frequency division multiplexers.

Keywords: resonances; periodic waveguides; reflection phases; topological properties

1. Introduction

In recent years, with the development of ultra-fast technologies, the research on terahertz (THz) technology has seen unprecedented progress [1–4]. THz waves have many unique advantages over the electromagnetic waves in the other bands, which allow THz technology to have considerable applications in some important areas, such as military, astronomy, radar, and medicine [5–8]. Due to its strong penetrability and low photonic energy, THz imaging technology will replace X-rays in medical examinations and will not cause harm to the human body [9]. It can also be used for detection in complex environments and identification of plastic weapons [10]. The characteristic spectra of many biomolecules and chemicals are also located in the THz band, including many drugs. So, it is also useful to employ THz waves in substance identification and biomedical research [11]. The broadband characteristics of THz waves result in their predictable applications in communication fields, and their wireless transmission speed can reach tens of Gb/s [12,13]. In the upcoming 6G Internet of Things, THz waves are also considered to be good candidates in short-range, bandwidth-aggressive services such as in the smart home scenario [14]. In THz systems, functional devices are essential to applications, including, of cause, displacement sensors. For example, in THz imaging, displacement sensing can correct the errors of relative motion between objects and probes, and it is expected to be adopted in calibrating the amplitude of characteristic peaks for different substances. In this era of interconnection of all things, most communications depend on THz waves, and displacement sensors can be integrated into smart devices to monitor and control their position changes.

The micro–nano displacement sensor as a device for high-precision monitoring has been used in automotive industry [15], small-scale manipulation [16,17], construction [18,19], micro-grippers [20], physiological sensing [21,22], and so on. The micro–nano displacement sensing technology of the communication optical band has been developed over many years, including plasmonic slot metamaterials [23], Fabry–Pérot interferometers [24], photonic crystal fibers [25], and so on. In 2011, Liu et al. reported a sensing structure with double-fiber Bragg gratings [26]. In 2014, Qu et al. presented an interferometric fiber-optic bending/micro-displacement sensor based on a plastic dual-core fiber with one end coated by a silver mirror [27]. In 2016, Gao et al. realized an optical displacement sensor based on anti-resonant reflecting guidance in a capillary covered hollow-core fiber [28]. However, a THz micro–nano displacement sensor of waveguide types that can be applied in THz systems without additional optical devices has not been reported yet. It will be very intriguing to realize THz micro–nano displacement sensing with high precision, small size, and easy integration. Hypersensitive THz displacement sensors could expand the applications of THz technology and pave the way for THz micro–nano positioning in the future smart life of all things connected.

Here, we propose a THz displacement sensor based on the hetero-structure waveguide, which can provide a sensitive spectrum response to a tiny displacement. Hetero-structures consist of semiconductor materials with different forbidden bands [29–31], which have been developed for artificial structures for years [32–36]. Further, the waveguides in the THz frequency range have been investigated numerically for functional devices [37,38]. The proposed hetero-structure contains a cylindrical waveguide with two different periodic corrugations on the wall. Based on their different topological properties, these two periodicities are used to generate an interface state, which can be recognized by a very narrow transmitted peak in its spectrum. It was found that the transmitted peak highly depends on the cavity length at the interface of two periodicities. Increasing the cavity length can result in red shift of the peak, which is very promising for use in fabricating a sensitive THz sensor. In the following section, the hetero-structure waveguide and its arising interface state are demonstrated. There is a transmitted peak arising in the former forbidden band due to the interface of the two periodic structures. The hetero-structure waveguide composed of two opaque periodic tubes become transparent at a certain frequency. The different topological properties of the two periodic structures can be explained by their reflection phases at the interface. The displacement THz sensor is proposed in Section 3 and the sensing performance is discussed in detail. Finally, the major findings about the super-high resolution and sensitivity are summarized.

2. Hetero-Structure Waveguides

Hetero-structure waveguides are usually combinations of multiple different tubes with unequal topological band gaps. Between each waveguide of the hetero-structures, interface states can always arise to produce an extraordinary transmitted peak in the former forbidden bands of periodic waveguides, which is very promising for high-resolution sensing due to its very narrow line width. Here, we propose a very simple corrugated waveguide system to demonstrate a THz micro–nano displacement sensor with very high precision. In waveguides with periodic wall corrugations, frequency gaps will appear around the resonant frequencies, in which the electromagnetic waves cannot pass through the structures. Bragg resonance will happen when the longitudinal wavenumbers of the same transverse modes satisfy the matching condition [39]. Connecting two periodic waveguides with similar Bragg gaps can result in an interface state with a very sharp transmitted peak in the overlapping frequency band gap when their topologies are different. These two waveguides have the same average radius r_0 and period Λ, the transmission spectra and geometry structures of which are shown in Figure 1a–c. The duty ratio refers to the proportion of wide radius part in a period length, and it has been proved that different duty ratios can introduce different topological band gaps [40]. We selected different duty ratios 0.8 and 0.4 for Waveguides A and B, respectively, to achieve the different topologies.

In a hollow metallic cylindrical periodic waveguide, the lowest-order Bragg resonance happens at

$$f_r = \frac{c}{2\pi}\sqrt{\frac{k_r^2}{r_0^2} + \left(\frac{\pi}{\Lambda}\right)^2} \quad (1)$$

for transverse magnetic (TM) waves with k_r = 2.4048. To set the Bragg resonance around 1 THz, we selected the geometry parameters of the periodic waveguides as R_{I} = 220 µm, R_{II} = 180 µm, and Λ = 183 µm with $r_0 = (R_{\mathrm{I}} + R_{\mathrm{II}})/2$ = 200 µm, where R_{I} and R_{II} are the wide and narrow radii of the structures, respectively. The hetero-structure waveguide was obtained by directly connecting the wide tube of Waveguide A and the narrow tube of Waveguide B.

We employed finite element method simulations on the waveguide structures with an axisymmetric model in COMSOL Multiphysics software. In the simulations, the refractive index of inside air was set to be 1, and perfect electrical boundary conditions were assigned to the walls of these three waveguides. As the THz source, a first-order TM mode was excited at the inlet of each waveguide, and the radiation condition was assigned to the outlet boundary. The 0.8–1.1 THz frequency range was selected with a step of 0.001 THz, and the electromagnetic fields were calculated in the whole waveguide.

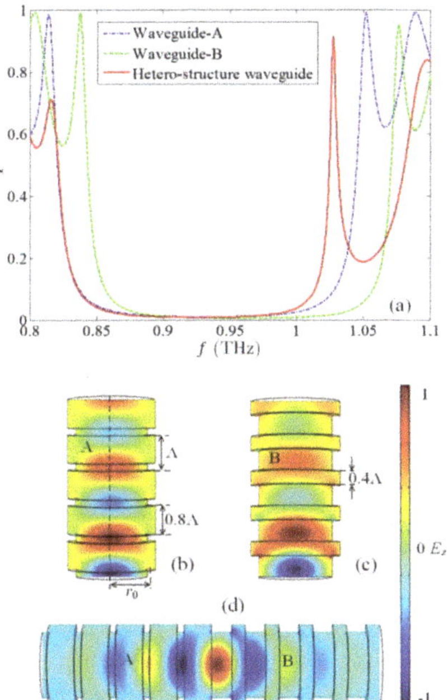

Figure 1. Transmission spectrum and electric field E_z components of Waveguides A, B, and the hetero-structure of their combination. (**a**) The blue dash–dot, green dash, and red solid lines are the transmission spectra of Waveguides A and B and the hetero-structure, respectively. The electric fields E_z of Waveguides A (**b**) and B (**c**) and the hetero-structure (**d**) at 1.027 THz illustrate the energy attenuation and localization at the interface of the structure. The electric fields were normalized by their own maximum in each waveguide.

Defining the transmission coefficient T as the ratio of optical powers of the outlet and inlet, we calculated the transmission spectra of Waveguides A and B and the hetero-structure waveguide, and depicted them by the dash–dot, dashed, and solid lines in Figure 1a. It shows that Waveguide A has a band gap around 0.838–1.038 THz, and Waveguide B's is around 0.848–1.051 THz. The smaller duty ratio of Waveguide B leads to a small frequency shift of the whole Bragg gap to a higher frequency. However, these two waveguides have a common forbidden band in the frequency range of 0.848–1.038 THz, in which the THz radiation cannot pass through either Waveguide A or B. The frequency of 1.027 THz falls in this common forbidden band—that is to say, a THz wave at this frequency cannot pass through either of these two waveguides—but what a surprise that we find an extraordinary transmitted peak with a center frequency of 1.027 THz in the spectrum of the hetero-structure waveguide. The forbidden band becomes much wider than that of either of the two waveguides.

To verify the interface state arising, we also simulated the electric field distributions for these three different waveguides, and the E_z components of the electric fields at 1.027 THz are depicted in Figure 1b–d for Waveguides A and B and the hetero-structure, respectively. The electric fields were normalized by their own maxima in the figure. The THz waves enter the waveguide from the bottoms of Waveguides A and B in Figure 1b,c respectively. It can be observed that the THz wave decays along the direction of propagation. Fortunately, the frequency is very close to the edge of the forbidden bands. So, we find that Waveguide B with a smaller duty ratio is much more effective for THz attenuation. In any case, the THz radiation at 1.027 THz cannot pass through either Waveguide A or B. When we connect these two waveguides and excite a THz wave with the first TM mode at the left side of the hetero-structure waveguide, as shown in Figure 1d, the situation turns out to be totally different. The THz waves accumulate around the interface of Waveguides A and B, which is known as the interface state. Due to the energy accumulation, the former opaque waveguides become transparent to a special frequency THz wave, as shown in Figure 1a. It is because of the normalization to the maximum electric field that the E_z component at the outlet is smaller than 1. Based on the transmission spectrum and the electric field distribution, we confirm the interface state arising in the proposed hetero-structure waveguide.

The remarkable transmission feature is due to the different topological properties of the two waveguides, which can be identified by the Zak phases of the two Bragg bands [40]. It is also more convenient to investigate the reflection phase of each waveguide at the interface. If the phases have the same sign, the related Bragg gaps are of a similar topology. Otherwise, they are topologically different, and the interface state arises.

To achieve the reflection phase of each waveguide, we connected a straight tube to the corrugated one and simulated the E_z component of electric fields with an incident TM mode from the straight tube at 1.027 THz. A straight tube with a length of 1000 μm and radius of 200 μm was connected to the wide radius port of Waveguide A. For Waveguide B, it was connected to the narrow radius port. Thus, the reflection properties at the interface of each waveguide could be observed. The same straight tube with a perfect electrical boundary at the other end was also simulated for reference. The reflected components of the electric fields are shown in Figure 2a–c for Waveguide A, the perfect electrical boundary, and Waveguide B, respectively. For convenience, we also depict the amplitude of E_z by solid lines. With the aid of the added dashed line, we can conclude that the reflection phase of Waveguide A is delayed while that of Waveguide B is advanced when considering the perfect electrical boundary reference as a zero-phase case. The reflection phases of Waveguides A and B have opposite signs, indicating the topological difference of the two Bragg gaps. So, the interface state arises at 1.027 THz, where we found a narrow transmitted peak in the spectrum. Such an extraordinary peak with very narrow line width would be a good candidate for THz sensing applications.

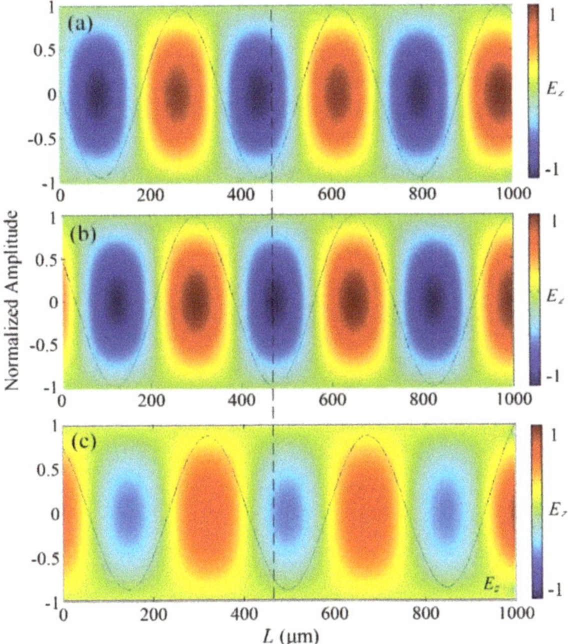

Figure 2. E_z components of reflected electric fields at 1.027 THz for (**a**) Waveguide A, (**b**) the perfect electrical boundary, and (**c**) Waveguide B. The reflected phase of Waveguide A is delayed while that of Waveguide B is advanced relative to the THz waves reflected by the perfect electrical boundary.

3. Micro–Nano Displacement Sensing

To realize micro–nano displacement sensing, we propose the hetero-structure waveguide (Figure 3a) composed of two tubes with average inner radius 200 μm and period 183 μm based on the above analysis. The tube wall is suggested to be 10 μm thick silver (Ag), which has low loss in the THz frequency range. So, the outer radii are 190 μm and 230 μm for the narrow and wide parts, respectively. According to the duty ratios, the lengths of the wide and narrow tubes are 146.4 μm and 36.6 μm, respectively, for Waveguide A, whereas they are 73.2 μm and 109.8 μm for Waveguide B. To fabricate an integrated device, we have to increase the length of the wide tube at the right end of Waveguide A and add a ring at the left end of Waveguide B. The increased length is 276 μm and the outer radius of the ring should be 220 μm to make sure that they can be connected. The length can be mechanically changed within the range of 20–120 μm. Although the measurement cannot start from 0 μm due to the length of the narrow tube in Waveguide B, the measurement range of displacement can still reach 100 μm. The number of periods in each waveguide is 5.

We also performed simulations on different combinations of duty ratios. The duty ratios of Waveguides A and B were selected from 0.1 to 0.9 in intervals of 0.1. The results show that all connections of different duty ratios can create a similar extraordinary transmitted peak, that is to say, the interface state between the two periodic waveguides always arises in the former Bragg gaps due to the different topologies. However, the bandwidth and frequency shift of the peaks highly rely on the duty ratios. Only the duty ratio combination of 0.8 and 0.4 can result in the narrowest bandwidth and maximum displacement, which could be of extreme benefit for practical applications. So, the proposed structure with duty ratios 0.8 and 0.4 was analyzed in detail for THz displacement sensing.

In the displacement sensing, we first fix Waveguide A, then adjust the Waveguide B to a relative position to be measured, and finally fix Waveguide B to that position. In this way, Waveguide B moves relative to Waveguide A, which increases the length of the cavity between the two waveguides. The THz wave is incident from the narrow radius of Waveguide A and emitted from the wide radius of Waveguide B. The frequency of the transmitted peak shifts with the displacement, so that the hetero-structure waveguide realizes displacement sensing. The simulated transmission for different displacements is depicted in Figure 3b by different lines. The center frequency of the transmitted peak is 1.002 THz when the displacement is 20 μm in the hetero-structure waveguide. When Waveguide B moves away from Waveguide A, the length of the middle cylindrical waveguide increases and the peak frequency shifts to the low frequency range. All the transmissions are above 0.85, and the highest transmission of 0.948 is obtained when the displacement is 120 μm.

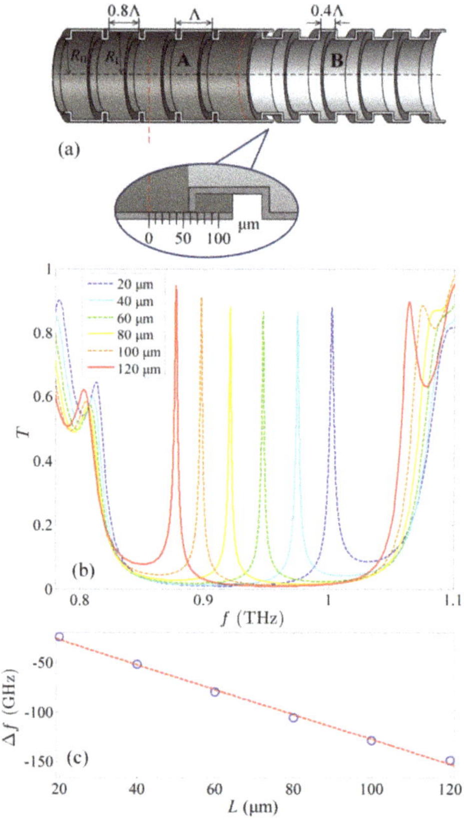

Figure 3. The THz micro–nano displacement sensor and its sensing performance. (**a**) Structure diagram of the hetero-structure waveguide sensor. The dark and light gray parts are Waveguides A and B, respectively. The red dashed line denotes the zero-displacement point and the inset is a magnification of the shifting structure with scales. (**b**) Transmitted peaks for the different displacements according to the scales. (**c**) Peak frequency shifting with increasing displacement.

To measure the displacement by THz waves, we define the frequency shift Δf from the frequency of 1.026 THz at 0 μm. The former peak moves from 1.027 THz to 1.026 THz for the 0 μm waveguide because the real dielectric constant of Ag is considered in the simulations. The frequency shift Δf according to the displacement is marked in Figure 3c by the circles, and we performed a linear fit (the dashed line) as follows:

$$\Delta f = -1.260 \times L \tag{2}$$

where L is the displacement in micrometers, and the frequency shift is in gigahertz. The fitting results show that when the hetero-structure waveguide is stretched 1 μm, the transmitted peak moves 1.260 GHz to a lower frequency. When L = 80 μm, the narrowest bandwidth is 2.580 GHz, and when L = 20 μm, the maximum bandwidth is 4.610 GHz. The maximum displacement that can be measured by the hetero-structure waveguide is 100 μm, the adjustable range of frequency is 0.877–1.002 THz, and the whole bandwidth is 124.5 GHz. So, the sensitivity of such a waveguide-type sensor can reach over 1.2 GHz/μm, and the minimum resolvable length is 2.073 μm when the displacement L is around 80 μm.

To realize the proposed structure, there are two ideas for fabrication. The first idea is to make a hollow metallic waveguide with substrates outside, as shown in Figure 4a. The deep lithography process can be employed to machine two half-cylinder polymers, and the metallic materials, such as Ag, can be sputtered on the two parts. To form the tube, we can hold the two parts together and eliminate a thin layer at the end of Waveguide B. The second idea is to produce a corrugated polymer core by 3D printing, as shown in Figure 4b. The two structures both end at the narrow radius parts. A silver film can be coated on the surface of the corrugated core, and a capillary coated with a silver film inside can be used to connect the two waveguides. Thus, the shifting parts are still hollow. When the proposed structure is ready, Waveguides A and B can be fixed to the two holders using epoxy glue [28]. With the aid of a 3D nano-positioning stage or optical micromanipulation, we can assemble the sensor to the test structures or THz systems.

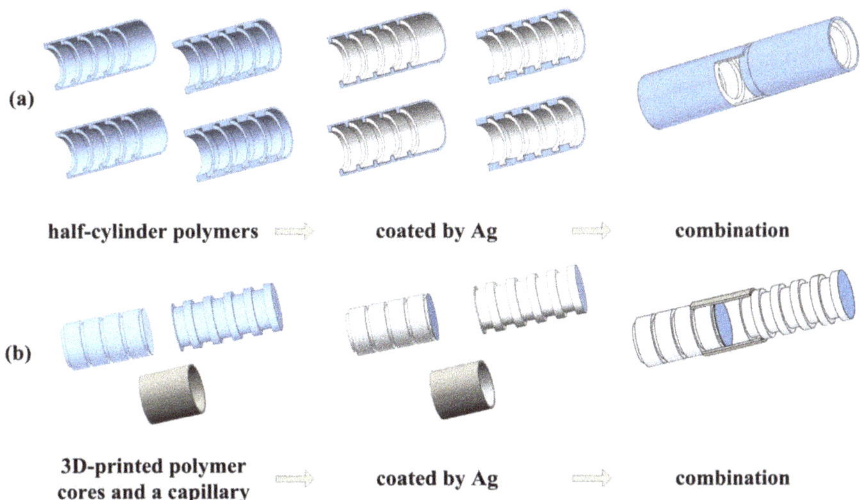

Figure 4. Sketch maps of the device fabrication process with the substrates outside (**a**) or inside (**b**).

The number of periods is also an important parameter for the hetero-structure waveguide sensor and can directly affect its resolution. Here we studied the effects of the number of periods on the sensing performance. The number of periods of the two waveguides takes the same value and is denoted by N. The value of N ranges from 4 to 9, while the length L is fixed at 80 μm. The transmission spectra and the measurement resolutions of the micro–nano displacement sensor are shown in Figure 5a,b, respectively, in the cases of different N. The bandwidth becomes extremely narrow as the number of periods increases, but when the number of periods is greater than 8, the transmission gets a little bit smaller. When N = 9, the transmission is just over 0.2. When the number of periods is smaller than 7, the transmissions are all greater than 0.5. The variation of the measurement resolution according to the number of periods N for the hetero-structure sensor is also depicted in Figure 5b by the crosses, with its fitting curve (the solid line) as

$$Rs = 7063 \times N^{-5.083} \tag{3}$$

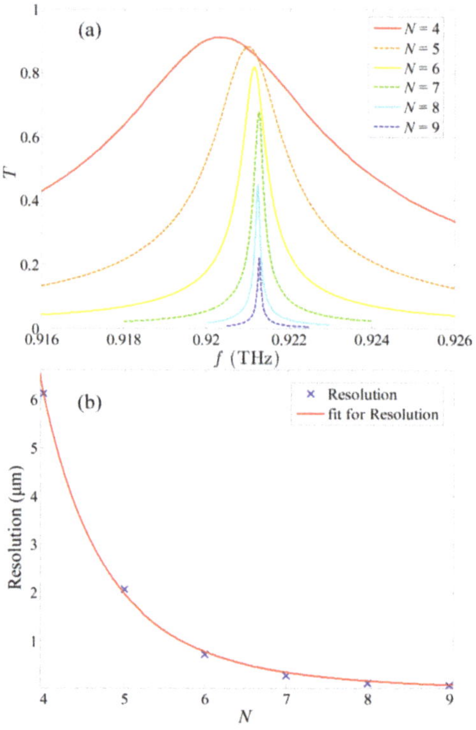

Figure 5. Transmitted peaks (**a**) and measurement resolutions (**b**) for different numbers of periods.

The resolution of the sensor is defined as its smallest measurable displacement, which is inversely proportional to the fifth power of N. When increasing the number of periods, we can greatly improve the performance of the sensor. The resolution reaches 282 nm for N = 7, while it gets as small as 90 nm for N = 9. The related sensitivity is over 1.2 GHz/μm. This is a highly sensitive THz nano displacement sensor that can be used in various applications involving accurate displacement measurements. It can also be extended to a wider range of applications when cascade hetero-structure waveguides are considered.

Through simulation of the hetero-structure waveguide, it was proved that the structure can produce a very narrow THz pulse. By changing the connection length between Waveguides A and B, the frequency of the transmitted peak moves towards the low frequency range, thus determining the micro–nano displacement sensing function of the proposed structure in the THz region. For measuring car body paints, a THz transceiver was mounted on a robot [41]. The proposed sensor can be integrated to monitor tiny movements of the mechanical arms without any additional sources. Based on the recorded THz data, the extracted layer thickness can be corrected. In the Internet of Things, THz functional devices play a key role in communication and intelligence applications [42]. The integrated displacement sensor can not only provide more accurate locations of communication nodes, but also monitor subtle changes of smart terminals. The structure design and performance simulations confirmed THz displacement sensing based on the hetero-structure waveguide.

4. Conclusions

We proposed a micro–nano displacement sensor based on hetero-structure waveguides in the THz frequency range. It is composed of two periodically corrugated waveguides that have similar Bragg forbidden bands. The THz waves cannot pass through either of these two periodic waveguides in the common frequency range of 0.848–1.038 THz. The topological analysis indicates that the Bragg gaps can present different topological properties, which can create an interface state with a very sharp transmitted peak in the THz spectrum when these two periodic waveguides are connected. Due to the topological difference of the Bragg gaps, the hetero-structure waveguide turns transparent at a certain frequency, where both the waveguides are otherwise opaque. Based on the interface state's induced transmission, we proposed a THz micro–nano displacement sensor by carefully connecting two corrugated waveguides. When the two waveguides are held by different structures, their relative displacement can be achieved by measuring the shift of transmission peaks, and the resolution can be improved by increasing the number of periods. The proposed hetero-structure waveguide-type sensor has excellent performance, such as a wide measurable range of 100 μm, minimum resolution of 90 nm, and maximum sensitivity of over 1.2 GHz/μm, which allow it various applications in many fields, such as chemical and biomedical sensing, micro-manipulation, imaging, and intelligent control. Besides displacement sensing applications, the proposed interface state of the hetero-structure waveguide can also be applied to other functional devices, such as switches, filters, and frequency division multiplexers.

Author Contributions: Conceptualization, Y.-X.F. and Z.-Y.T.; methodology, H.L. and L.-L.X.; validation, L.-L.X., and H.L.; formal analysis, Y.-X.F., T.Z., and Z.-Y.T.; writing—original draft preparation, L.-L.X.; writing—review and editing, Y.-X.F. and Z.-Y.T.; visualization, L.-L.X.; supervision, Y.-X.F. and Z.-Y.T. All authors have read and agreed to the published version of the manuscript.

Funding: This research was funded by the Dean Project of Guangxi Key Laboratory of Wireless Broadband Communication and Signal Processing, the National Natural Science Foundation of China (11374071), and the Natural Science Foundation of Heilongjiang Province, China (A2018004).

Conflicts of Interest: The authors declare no conflict of interest.

References

1. Köhler, R.; Tredicucci, A.; Beltram, F.; Beere, H.E.; Linfield, E.H.; Davies, A.G.; Ritchie, D.A.; Iotti, R.C.; Rossi, F. Terahertz semiconductor-heterostructure laser. *Nature* **2002**, *417*, 156–159. [CrossRef] [PubMed]
2. Tonouchi, M. Cutting-edge terahertz technology. *Nat. Photon.* **2007**, *1*, 97–105. [CrossRef]
3. Thoma, P.; Scheuring, A.; Hofherr, M.; Wünsch, S.; Ilin, K.; Smale, N.; Judin, V.; Hiller, N.; Müller, A.S.; Semenov, A.; et al. Real-time measurement of picosecond THz pulses by an ultra-fast $YBa_2Cu_3O_{7-d}$ detection system. *Appl. Phys. Lett.* **2012**, *101*, 142601. [CrossRef]
4. Preu, S.; Mittendorff, M.; Winnerl, S.; Lu, H.; Gossard, A.C.; Weber, H.B. Ultra-fast transistor-based detectors for precise timing of near infrared and THz signals. *Opt. Express* **2013**, *21*, 17941–17950. [CrossRef] [PubMed]
5. Cazzoli, G.; Cludi, L.; Buffa, G.; Puzzarini, C. Precise THz measurements of HCO^+, N_2H^+, and CF^+ for astrophysical observations. *Astrophys. J. Suppl. Ser.* **2012**, *203*, 1–9. [CrossRef]
6. Yang, Q.; Qin, Y.; Deng, B.; Wang, H.; You, P. Micro-Doppler ambiguity resolution for wideband terahertz radar using intra-pulse interference. *Sensors* **2017**, *17*, 993. [CrossRef]
7. Jiang, Y.; Deng, B.; Wang, H.; Qin, Y.; Liu, K. An effective nonlinear phase compensation method for FMCW terahertz radar. *IEEE Photon. Technol. Lett.* **2016**, *28*, 1684–1687. [CrossRef]
8. Jiang, X.; Chen, H.; Li, Z.; Yuan, H.; Cao, L.; Luo, Z.; Zhang, K.; Zhang, Z.; Wen, Z.; Zhu, L.; et al. All-dielectric metalens for terahertz wave imaging. *Opt. Express* **2018**, *26*, 14132–14142. [CrossRef]
9. Knobloch, P.; Schildknecht, C.; Kleine-Ostmann, T.; Koch, M.; Hoffmann, S.; Hofmann, M.; Rehberg, E.; Sperling, M.; Donhuijsen, K.; Hein, G.; et al. Medical THz imaging: An investigation of histo-pathological samples. *Phys. Med. Biol.* **2002**, *47*, 3875–3884. [CrossRef]

10. Federici, J.F.; Schulkin, B.; Huang, F.; Gary, D.; Barat, R.; Oliveira, F.; Zimdars, D. THz imaging and sensing for security applications-Explosives, weapons and drugs. *Semicond. Sci. Technol.* **2005**, *20*, S266–S280. [CrossRef]
11. Fischer, B.M.; Walther, M.; Jepsen, P.U. Far-infrared vibrational modes of DNA components studied by terahertz time-domain spectroscopy. *Phys. Med. Biol.* **2002**, *47*, 3807–3814. [CrossRef] [PubMed]
12. Ferraro, A.; Tanga, A.A.; Zografopoulos, D.C.; Messina, G.C.; Ortolani, M.; Beccherelli, R. Guided mode resonance flat-top bandpass filter for terahertz telecom applications. *Opt. Lett.* **2019**, *44*, 4239–4242. [CrossRef] [PubMed]
13. Wang, C.; Lu, B.; Lin, C.; Chen, Q.; Miao, L.; Deng, X.; Zhang, J. 0.34-THz Wireless Link Based on High-Order Area Network Applications. *IEEE Trans. Terahz. Sci. Technol.* **2014**, *4*, 75–85. [CrossRef]
14. Mao, B.; Kawamoto, Y.; Kato, N. AI-based Joint Optimization of QoS and Security for 6G Energy Harvesting Internet of Things. *IEEE Internet Things J.* **2020**. [CrossRef]
15. Kim, Y.S.; Choi, Y.H.; Lee, J.M.; Noh, J.S.; Kim, J.J.; Bien, F. Fabrication of a novel contactless switch using eddy current displacement sensor for safer vehicle brake system. *IEEE Trans. Veh. Technol.* **2011**, *60*, 1485–1495.
16. Li, Y.; Xu, Q. Adaptive sliding mode control with perturbation estimation and PID sliding surface for motion tracking of a piezo-driven micromanipulator. *IEEE Trans. Control Syst. Technol.* **2010**, *18*, 798–810. [CrossRef]
17. Liaw, H.C.; Shirinzadeh, B. Constrained motion tracking control of piezo-actuated flexure-based four-bar mech- anisms for micro/nano manipulation. *IEEE Trans. Autom. Sci. Eng.* **2010**, *7*, 699–705. [CrossRef]
18. Kanekawa, K.; Matsuya, I.; Sato, M.; Tomishi, R.; Takahashi, M.; Miura, S.; Suzuki, Y.; Hatada, T.; Katamura, R.; Nitta, Y.; et al. An experimental study on relative displacement sensing using phototransistor array for building structures. *IEEJ Trans. Electr. Electron. Eng.* **2010**, *5*, 251–255. [CrossRef]
19. Yu, H.; He, X.; Ding, W.; Hu, Y.; Yang, D.; Lu, S.; Wu, C.; Zou, H.; Liu, R.; Lu, C.; et al. A Self-Powered Dynamic Displacement Monitoring System Based on Triboelectric Accelerometer. *Adv. Energy Mater.* **2017**, *7*, 1700565. [CrossRef]
20. MacKay, R.E.; Le, H.R.; Clark, S.; Williams, J.A. Polymer micro-grippers with an integrated force sensor for biological manipulation. *J. Micromech. Microeng.* **2013**, *23*, 015005. [CrossRef]
21. Kang, D.; Pikhitsa, P.V.; Choi, Y.W.; Lee, C.; Shin, S.S.; Piao, L.; Park, B.; Suh, K.Y.; Kim, T.; Choi, M. Ultrasensitive mechanical crack-based sensor inspired by the spider sensory system. *Nature* **2014**, *516*, 222–226. [CrossRef] [PubMed]
22. Wankhar, S.; Kota, A.A.; Selvaraj, D. A versatile stretch sensor for measuring physiological movement using a centre loaded, end-supported load cell. *J. Med. Eng. Technol.* **2017**, *41*, 406–414. [CrossRef] [PubMed]
23. Wu, W.; Ren, M.; Pi, B.; Cai, W.; Xu, J. Displacement sensor based on plasmonic slot metamaterials. *Appl. Phys. Lett.* **2016**, *108*, 073106. [CrossRef]
24. Zhu, C.; Chen, Y.; Du, Y.; Zhuang, Y.; Liu, F.; Gerald, R.E.; Huang, J. A Displacement Sensor with Centimeter Dynamic Range and Submicrometer Resolution Based on an Optical Interferometer. *IEEE Sens. J.* **2017**, *17*, 5523–5528. [CrossRef]
25. Dash, J.N.; Jha, R.; Villatoro, J.; Dass, S. Nano-displacement sensor based on photonic crystal fiber modal interferometer. *Opt. Lett.* **2015**, *40*, 467–470. [CrossRef]
26. Liu, F.; Fei, Y.; Xia, H.; Chen, L. A new micro/nano displacement measurement method based on a double-fiber Bragg grating (FBG) sensing structure. *Meas. Sci. Technol.* **2012**, *23*, 054002. [CrossRef]
27. Qu, H.; Yan, G.F.; Skorobogatiy, M. Interferometric fiber-optic bending/nano-displacement sensor using plastic dual-core fiber. *Opt. Lett.* **2014**, *39*, 4835–4838. [CrossRef] [PubMed]
28. Gao, R.; Lu, D.F.; Cheng, J.; Jiang, Y.; Jiang, L.; Qi, Z.M. Optical Displacement Sensor in a Capillary Covered Hollow Core Fiber Based on Anti-Resonant Reflecting Guidance. *IEEE J. Sel. Top. Quantum Electron.* **2017**, *23*, 5600106. [CrossRef]
29. Van de Walle, C.G.; Martin, R.M. Theoretical study of band offsets at semiconductor interfaces. *Phys. Rev. B* **1987**, *35*, 8154–8165. [CrossRef]
30. Tomozawa, H.; Numata, K.; Hasegawa, H. Interface states at lattice-matched and pseudomorphic heterostructures, *Appl. Surf. Sci.* **1992**, *60–61*, 721–728. [CrossRef]
31. Huang, Y.S.; Westenhoff, S.; Avilov, I.; Sreearunothai, P.; Hodgkiss, J.M.; Deleener, C.; Friend, R.H.; Beljonne, D. Electronic structures of interfacial states formed at polymeric semiconductor heterojunctions. *Nat. Mater.* **2008**, *7*, 483–489. [CrossRef] [PubMed]
32. Song, B.S.; Noda, S.; Asano, T. Photonic devices based on in-plane hetero photonic crystals. *Science* **2003**, *300*, 1537. [CrossRef]

33. Rechtsman, M.; Zeuner, J.; Plotnik, Y.; Lumer, Y.; Podolsky, D.; Dreisow, F.; Nolte, S.; Segev, M.; Szameit, A. Photonic Floquet topological insulators. *Nature* **2013**, *496*, 196–200. [CrossRef] [PubMed]
34. Fan, Y.-X.; Sang, T.-Q.; Liu, T.; Xu, L.-L.; Tao, Z.-Y. Single-mode interface states in heterostructure waveguides with Bragg and non-Bragg gaps. *Sci. Rep.* **2017**, *7*, 44381. [CrossRef] [PubMed]
35. Kim, I.; Iwamoto, S.; Arakawa, Y. Topologically protected elastic waves in one-dimensional phononic crystals of continuous media. *Appl. Phys. Exp.* **2018**, *11*, 017201. [CrossRef]
36. Tao, Z.-Y.; Liu, T.; Zhang, C.; Fan, Y.-X. Acoustic extraordinary transmission manipulation based on proximity effects of heterojunctions. *Sci. Rep.* **2019**, *9*, 1080. [CrossRef]
37. Gric, T.; Cada, M. Analytic solution to field distribution in one-dimensional inhomogeneous media. *Opt. Commun.* **2014**, *322*, 183–187. [CrossRef]
38. Gric, T.; Gorodetsky, A.; Trofimov, A.; Rafailov, E. Tunable plasmonic properties and absorption enhancement in terahertz photoconductive antenna based on optimized plasmonic nanostructures. *J. Infrared Millim. Terahz. Waves* **2018**, *39*, 1028–1038. [CrossRef]
39. Zhang, L.; Fan, Y.-X.; Liu, H.; Xu, L.-L.; Xue, J.-L.; Tao, Z.-Y. Hypersensitive and Tunable Terahertz Wave Switch Based on Non-Bragg Structures Filled with Liquid Crystals. *J. Light. Technol.* **2017**, *35*, 3092–3098. [CrossRef]
40. Xiao, M.; Ma, G.; Yang, Z.; Sheng, P.; Zhang, Z.Q.; Chan, C.T. Geometric phase and band inversion in periodic acoustic systems. *Nat. Phys.* **2015**, *11*, 240–244. [CrossRef]
41. Ellrich, F.; Bauer, M.; Schreiner, N.; Keil, A.; Pfeiffer, T.; Klier, J.; Weber, S.; Jonuscheit, J.; Friederich, F.; Molter, D. Terahertz Quality Inspection for Automotive and Aviation Industries. *J. Infrared Millim. Terahz. Waves* **2020**, *41*, 470–489. [CrossRef]
42. Hassan, N.; Chou, C.T.; Hassan, M. eNEUTRAL IoNT: Energy-neutral event monitoring for internet of nano things. *IEEE Internet Things J.* **2019**, *6*, 2379–2389. [CrossRef]

Article

On-Chip Terahertz Detector Designed with Inset-Feed Rectangular Patch Antenna and Catadioptric Lens

Fan Zhao [1], Luhong Mao [1], Weilian Guo [2], Sheng Xie [2,*] and Clarence Augustine T. H. Tee [3]

1 Department of Microelectronics and Solid-state Electronics, School of Electrical and Information Engineering, Tianjin University, Tianjin 300072, China; fanhere@tju.edu.cn (F.Z); lhmao@tju.edu.cn (L.M.)
2 Department of Microelectronics and Solid-state Electronics, School of Microelectronics, Tianjin University, Tianjin 300072, China; 201204056@tju.edu.cn
3 Department of Electrical Engineering, Faculty of Engineering, University of Malaya, Kuala Lumpur 50603, Malaysia; catht@um.edu.my
* Correspondence: xie_sheng06@tju.edu.cn; Tel.: +86-131-3223-3919

Received: 6 May 2020; Accepted: 19 June 2020; Published: 24 June 2020

Abstract: This study proposes an on-chip terahertz (THz) detector designed with on-chip inset-feed rectangular patch antenna and catadioptric lens. The detector incorporates a dual antenna and dual NMOSFET structure. Radiation efficiency of the antenna reached 89.4% with 6.89 dB gain by optimizing the antenna inset-feed and micro-strip line sizes. Simulated impedance was 85.55 − j19.81 Ω, and the impedance of the antenna with the ZEONEX horn-like catadioptric lens was 117.03 − j20.28 Ω. Maximum analyzed gain of two on-chip antennas with catadioptric lens was 17.14 dB resonating at 267 GHz. Maximum experimental gain of two on-chip patch antennas was 4.5 dB at 260 GHz, increasing to 10.67 dB at 250 GHz with the catadioptric lens. The proposed on-chip rectangular inset-feed patch antenna has a simple structure, compatible with CMOS processing and easily implemented. The horn-like catadioptric lens was integrated into the front end of the detector chip and hence is easily molded and manufactured, and it effectively reduced terahertz power absorption by the chip substrate. This greatly improved the detector responsivity and provided very high gain. Corresponding detector voltage responsivity with and without the lens was 95.67 kV/W with $NEP = 12.8$ pW/Hz$^{0.5}$ at 250 GHz, and 19.2 kV/W with $NEP = 67.2$ pW/Hz$^{0.5}$ at 260 GHz, respectively.

Keywords: THz detector; rectangular inset-feed patch antenna; catadioptric horn-like lens; CMOS process

1. Introduction

Terahertz (THz) detectors have been recently employed to assist security aspects for high-speed data communication, spectroscopy, concealed weapon detection, short-range radar, aviation assistance, cancer detection, and many more applications [1–4]. The THz spectrum region is in the middle of the microwave and infrared spectra and is sometimes known as sub-millimeter waves. Terahertz radiation ranges from 3 mm to 30 μm with corresponding frequencies of 0.1–10 THz [5–9], which penetrate plastics, paper, and wood. The THz region is sometimes called the "THz Gap" [10–12] because optical and microwave theory do not fully apply for this frequency range.

Specific terahertz antennas are required to radiate and receive terahertz waves, acting as transducers to convert high-frequency current or waveguide energy into spatial electromagnetic wave energy, and directional radiation for transmission or the reverse conversion for receiving. However, designing an on-chip antenna in the terahertz band is somewhat difficult, partly because the system's dynamic range and transmission distance between the THz wave source and detector are influenced by antenna gain. Dielectric layers between the CMOS top and bottom metal layers are too thin to construct a high gain antenna [13,14], and lossy silicon substrates significantly reduce on-chip patch antenna performance. However, problems can be solved with reasonable on-chip antenna design considering antenna gain and bandwidth.

Annular on-chip single-ended slot antennas can achieve approximately 4.5 dB gain with 50% radiation efficiency [15]. Although directional front-end antennas are unsuitable for THz detectors [16,17], Taeguk log periodic dipole antennas (LPDAs) can achieve a maximum 7.8 dBi gain at 73 GHz and circular LPDA antennas can achieve a maximum 9.25 dBi gain at 85 GHz [17,18]. The total size of these two antennas is 4.5×10 mm^2 and they have complicated structures. Rectangular antennas are the most common on-chip design, leveraging simple structure and being compatible with standard CMOS processing [18,19]. The proposed rectangular antenna from [20] was 237.5×317 μm^2 and operated at 300 GHz, but only achieved −2.1 dB gain. In contrast, the proposed THz dielectric resonator antenna (DRA) [21] employed a 500 μm resonator to greatly enhance antenna gain. The DRA was 290–310 μm wide and 390–410 μm long, considerably larger than the traditional rectangular patch antenna, and provided measured 6.7 dB maximum gain at 327 GHz. Thus, there is considerable ongoing research to improve THz detector antenna gain.

This study proposes an on-chip terahertz detector with on-chip rectangular inset-feed patch antennas that was implemented using a United Microelectronics Corporation (UMC) 0.18 μm standard CMOS process and ZEONEX RS420 (ZEON Corporation, Tokyo, Japan) horn-like catadioptric lens. Section 2 discusses the antenna design and Section 3 introduces the detector structure, combining plasma wave-based terahertz detection in NMOSFETs (TeraFETs). Section 4 describes the horn-like catadioptric lens and Section 5 details testing for the prototype proposed THz detector antenna and lens. Conclusions are presented in Section 6.

2. On-Chip THz Antenna

The UMC 0.18 μm CMOS process provides six metal layers with intermediate dielectric layers (SiO_2) between them. The CMOS process can be used to design not only digital and analog circuits, but also optoelectronic devices [22–24]. The major advantage is that signals detected by the detector can be amplified and processed on the chip. However, silicon is lossy in the THz frequency range due to the short wavelength. Therefore, we implemented the antenna inset-feed patch in the Metal 6 (top metal) layer, using the Metal 1 (bottom metal) layer as a ground plane to reflect radiated power back to the NMOSFETs, hence reducing THz signal penetration into the lossy silicon substrate [20,21]. Metal 6 is 2.06 μm thick and the SiO_2 CMOS process layer is 6.32 μm thick. Radiation efficiency was low because these layers were too thin to emit an EM field, e.g., a similar antenna design achieved only approximately 20% radiation efficiency in simulation [14,21], with −1.6 and 0.1 dB gains, respectively. In contrast, the proposed antenna achieved 89.4% simulated radiation efficiency and 6.89 dB simulated gain. Figure 1 shows a cross-section of the proposed CMOS structure and Figure 2 shows a suitable patch antenna design using UMC 0.18 μm CMOS technology. Patch height, dielectric substrate properties, and patch conductor thickness are fixed depending on the specific UMC technology employed. We adopted substrate thinning and increased the ground plane size to increase antenna gain, with a micro-strip feed line.

Rectangular patch antennas can fully meet CMOS process design rules, in contrast with circle, ring, and other irregularly shaped antennas. Harrington [25,26] provided analytical expressions to explain the relationship between antenna performance and this separation. The empirical relationships for rectangular patches can be expressed as follows [13,14]:

$$W = \frac{c}{2f_0}\sqrt{\frac{2}{\varepsilon_{SiO_2}}} \tag{1}$$

$$\varepsilon_{\text{eff}} = \frac{\varepsilon_{SiO_2}+1}{2} + \frac{\varepsilon_{SiO_2}-1}{2}\left(1+12\frac{h}{W}\right)^{-1/2} \tag{2}$$

$$\Delta L = 0.412h\frac{(\varepsilon_{\text{eff}} + 0.3)(W/h + 0.264)}{(\varepsilon_{\text{eff}} - 0.258)(W/h + 0.8)} \quad (3)$$

and

$$L = \frac{c}{2f_0\sqrt{\varepsilon_{\text{eff}}}} - 2\Delta L \quad (4)$$

where W is the radiation patch width, c is the speed of light in free space, f_0 is the operating frequency, ε_{eff} is the effective dielectric constant, h = 6.32 μm (for the UMC 0.18 μm CMOS process) is the dielectric thickness between the patch and ground, ε_{SiO_2} is the SiO_2 dielectric coefficient, and L is the patch length.

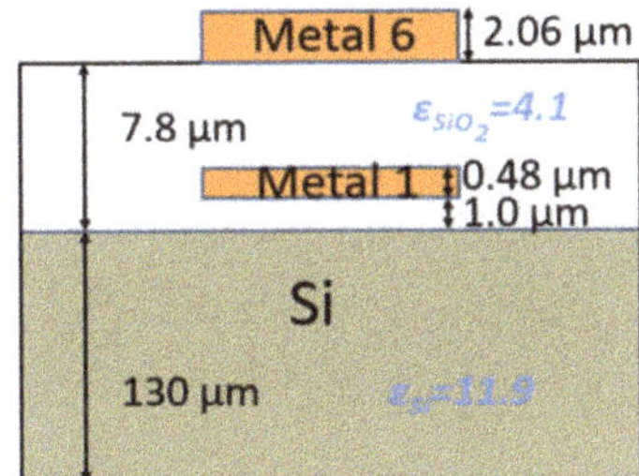

Figure 1. Cross-sectional view for the proposed United Microelectronics Corporation (UMC) 0.18 μm CMOS design.

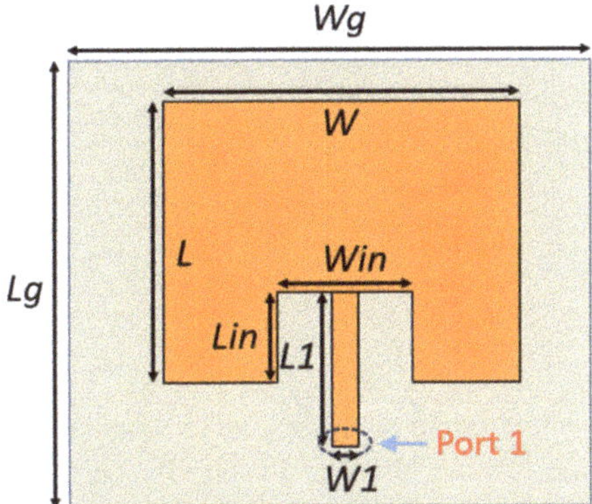

Figure 2. Structures for the proposed on-chip patch antenna.

Figure 2 shows the patch antenna structure, where *Wg* is the ground plane width, *Lg* is the ground plane length, *Win* is the insertion width, *Lin* is the insertion distance, *W1* is the micro-strip line width, and *L1* is the micro-strip line length.

Micro-strip and impedance matching networks were used to increase the input impedance. Table 1 shows the dimensions and performances of the patch antennas. Resonance is approximately 270 GHz and bandwidth is the region where $S11 < -10$ dB. Bandwidth for the patch antenna = 1.5 GHz with 271 GHz center frequency, and gain = 6.89 dB. Patch antenna (without lens) gain was higher than for other antennas, but still insufficient to provide a practical detector.

To ensure maximum power transmission, half the NMOSFET (TeraFET) channel impedance should be conjugated with the antenna output impedance. NMOSFET channel impedance depends on the gate voltage, hence the NMOSFETs should operate below the sub-threshold to obtain high impedance, forming a two-dimensional electronic fluid. The NMOSFET gate was powered by the antenna and matching network comprising a microstrip line and open stub line to achieve maximum power transmission. Performance comparison of terahertz antenna and lens are shown in Table 2, and without a lens or DRA, the patch antenna proposed in this study gets the highest gain of all.

Table 1. Proposed on-chip rectangular inset-feed patch antenna dimensions and performances [14].

Parameter	Patch
Wg (μm)	500
Lg (μm)	500
W (μm)	275
L (μm)	275
Win (μm)	90
Lin (μm)	65
W1 (μm)	5.5
L1 (μm)	122
Gain (dB)	6.89
Bandwidth (GHz)	1.5
Central frequency (GHz)	271
Radiation efficiency	89.4%

Table 2. Performance comparison.

Ref.	Process	Antenna	Center Frequency (GHz)	Gain (dB)	Bandwidth (GHz)
[27]	TSMC 65 nm CMOS	Ring	296	4.2	-
[28]	45 nm CMOS SOI	Slot ring	375	1.6	3
[21]	0.18 μm CMOS	Rectangle + DRA	327	7.9	2.5
[20]	TSMC 0.18 μm CMOS	Rectangle Circular Diamond	300 300 300	−2.1 5.30 5.94	4.1 11.3 9.6
[14]	0.18 μm CMOS	Rectangle	284	−1	7
[22]	65 nm CMOS	Ring with silicon lens	288	18.3	-
This work	UMC 0.18 μm CMOS	Rectangle	271	6.89	1.5

Antenna radiation efficiency η_{rad} is defined as the ratio of desired output power to supplied input power:

$$\eta_{rad} = \frac{P_{rad}}{P_{in}} = \frac{P_{in} - P_{loss}}{P_{in}} = 1 - \frac{P_{loss}}{P_{in}} \tag{5}$$

where P_{rad} is power radiated by the antenna, P_{in} is power supplied to the antenna input, and P_{loss} is power lost in the antenna. Other factors could also contribute to effective transmitted power loss, including impedance mismatch at antenna input, or receiving antenna polarization mismatch. However, these losses are external to the antenna and could be eliminated by proper matching networks, or proper receiving antenna choice and positioning. Therefore, these losses are not usually attributed to the antenna, compared with dissipative losses due to metal conductivity or dielectric loss within the antenna.

3. Terahertz Detector Circuit

Figure 3a shows the proposed detector structure. The antenna output port is connected to the matching network, and the network is connected to an NMOSFET (TeraFET) for improved power transfer efficiency. DC photoresponse ΔU of the on-chip amplifier output terminal was measured with an SR830 (Standard Research System, INC, Sunnyvale, CA, USA) lock-in amplifier which measured the ΔU amplitude submerged by noise at a determined frequency. Thus, the detector is thermal noise limited, and the detector also includes a dual antenna and dual NMOSFET (TeraFET) structure, which provides degree of noise suppression.

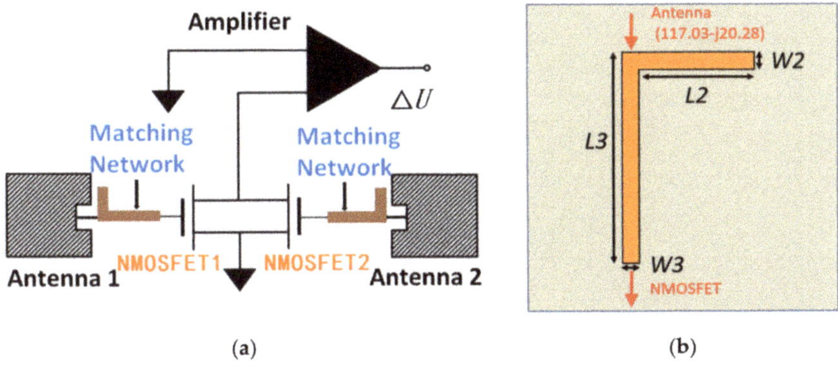

Figure 3. (**a**) MOSFET detector structure and (**b**) matching network between on-chip antenna and NMOSFET.

Antenna impedance is the ratio of input voltage to input current at the antenna feed. The best-case scenario is that antenna impedance is purely resistive and equal to half of the characteristic NMOSFET (TeraFET) impedance. There is no power reflection at the feeder terminal and no standing wave at the feed line, and antenna impedance changes slowly with frequency. Ideal matching will eliminate reactance. The NMOSFET's characteristic resistive impedance can be adjusted by increasing the applied voltage for better performance. Impedance matching between the antenna and NMOSFET (TeraFET) was achieved using an L-type matching network comprising a microstrip open-stub and microstrip line section to reduce the imaginary part magnitude as much as possible through the matching network, as shown in Figure 3b. Selected section dimensions were $L2$ = 105 μm, $W2$ = 13.6 μm, $L3$ = 222 μm, and $W3$ = 13.36 μm.

Figure 4 shows the circuit diagram for the proposed readout amplifier, comprising two stages. Stage 1 uses a common-source structure with large differential NMOS input pair (M1 and M2), since increased size can reduce input-referred flicker noise. M3 and M4 are also large to similarly reduce any contribution to flicker noise. M5 and M6 operate in their linear region, acting as resistors to provide common-mode feedback. C0 is a 10 pF capacitor with larger equivalent resistance at low frequency. M8 must operate in the sub-threshold region to provide direct current bias for M1 and M2 to ensure that the alternating current signal passes through the capacitor to M2 and M7 gate.

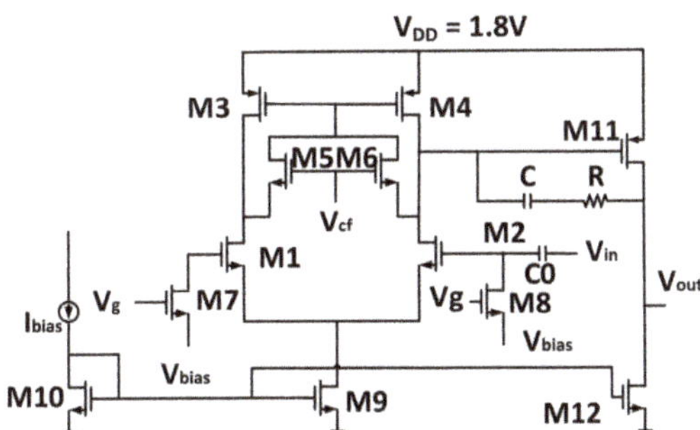

Figure 4. Proposed low-noise amplifier (LNA).

Stage 2 uses a PMOS input device with a common-source single-ended structure. The R and C provide Miller compensation for circuit stability when operated as a closed loop.

The proposed on-chip low-noise amplifier (LNA) shown in Figure 4 compensates for low-gain antennas. Two key issues must be solved during circuit design, namely low-noise performance and amplification ability under low frequency. The amplifier, fabricated in the standard 0.18 μm CMOS process, provides 42 dB voltage gain and 4 MHz 3 dB bandwidth with capacitor feedback. Simulated input-referred noise is 17 nV/Hz$^{1/2}$ at 10 kHz and 113 nV/Hz$^{1/2}$ at 310 Hz. Amplifier power consumption is 3.5 mW with 1.8 V supply voltage. Figure 5 shows the input-referred noise curve.

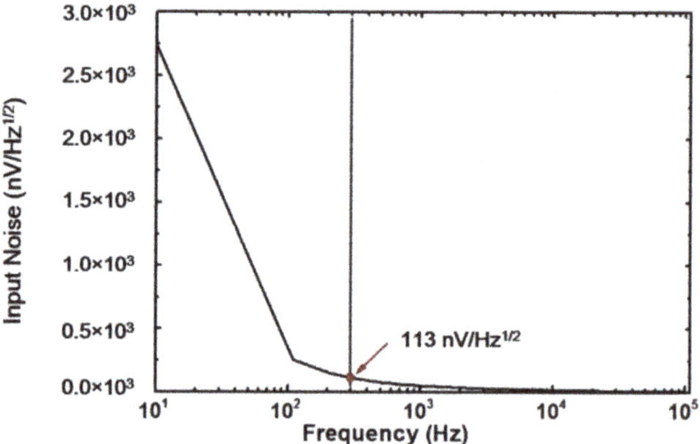

Figure 5. Low-noise amplifier (LNA)'s input-referred noise.

4. Catadioptric Horn-Like Lens

Traditional catadioptric optical systems are commonly used for optical telescopes, telephoto lenses, early lighthouse focusing systems, searchlights, etc., with refraction achieved by lenses and reflection by curved mirrors. This lens can be easily integrated with THz detectors in contrast with conventional quasi-optical components, as shown in Figure 6. Terahertz detector front-end antennas have been physically coupled to back-end silicon lens previously to yield a compact system [10]. Therefore, this study proposes a THz detector with front-end antenna and lens symmetric about the z-axis to reduce signal loss in the substrate.

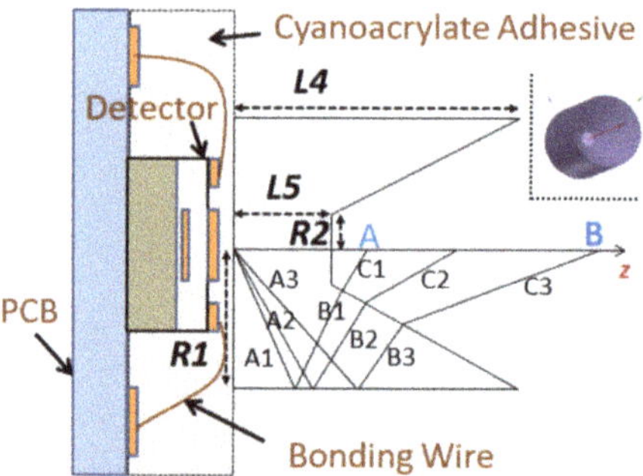

Figure 6. Cross-sectional view for the proposed catadioptric lens with inset 3D view of the lens.

Figure 7 shows a simulated lens, where the lens' out-radius $R1$ = 3 mm and in-radius $R2$ = 0.6 mm ($R1$ is about $\varepsilon\lambda_0$, where λ_0 is the free-space wavelength [29]). Catadioptric lens length $L4$ = 6 mm, corresponding to $2\varepsilon\lambda_0$ at 270 GHz. Length between out-radius and in-radius $L5$ = 2 mm, approximately λ_0. When A1, A2, and A3 rays radiate from the source, they are totally reflected at the reflection plane (B1, B2, and B3, respectively), which are then refracted at the refraction plane (C1, C2, and C3, respectively).

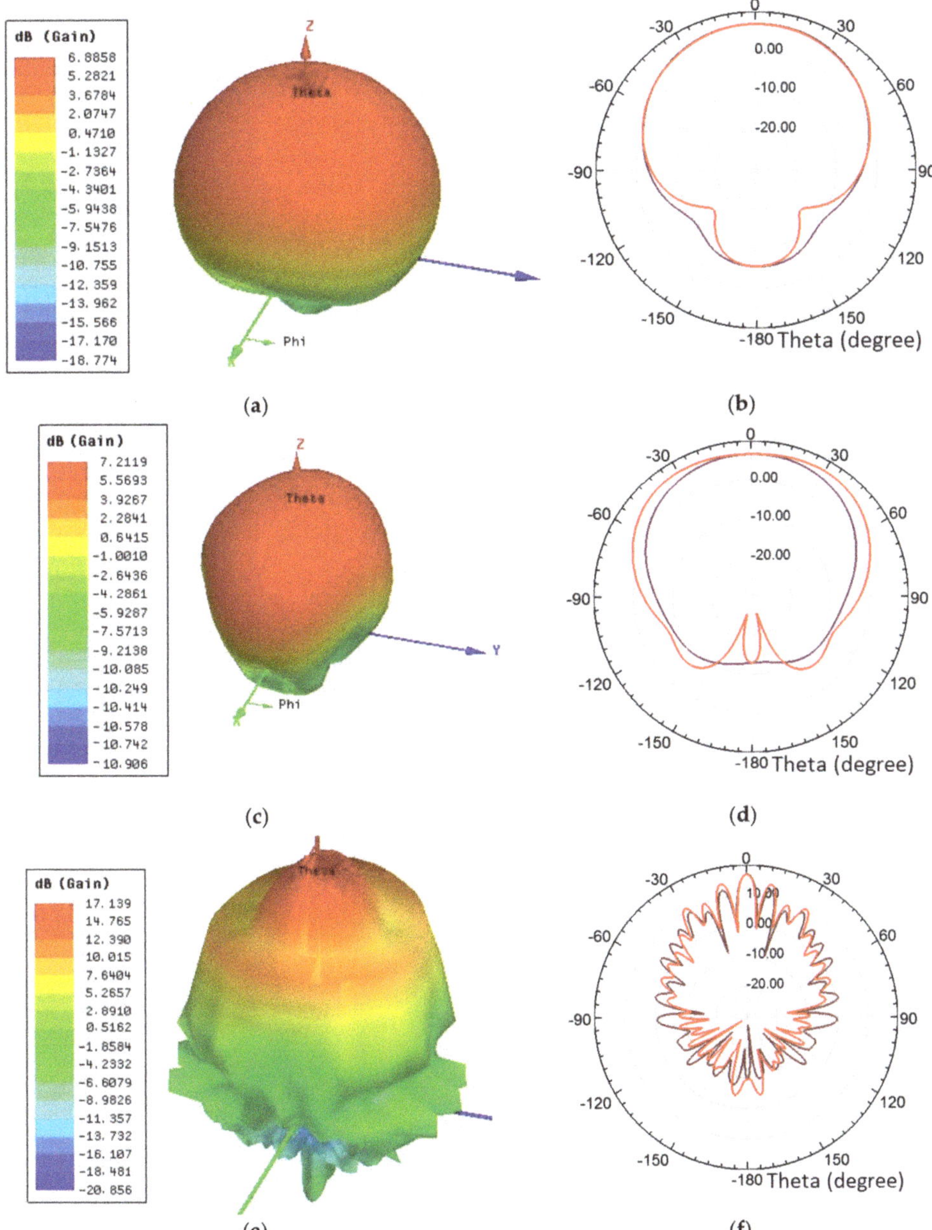

Figure 7. Simulated gain at 280 GHz (**a**,**b**) individual patch antenna, (**c**,**d**) two patch antennas working together, and (**e**,**f**) two patch antennas with catadioptric lens. Red and brown lines represent 2D simulated radiation pattern along Phi = 0° and 90° planes, respectively.

The THz wave power received by the antenna is proportional to the antenna area, hence the two antennas receive approximately twice the power as a single antenna. The signal is converted to a direct voltage signal by the detector. Although the two NMOSFETs (TeraFETs) are connected in parallel, the output voltage does not increase but the output current doubles. Since the equivalent impedance for the low-noise amplifier input is large, the output current is converted into a voltage, hence doubling the voltage signal after passing through the low-noise amplifier.

Figures 6 and 7 show that the refracted rays are distributed and converged to the z-axis, and the distribution provides up to 17.14 dB total gain, considerably higher than that achieved without the lens. The catadioptric dielectric lens is easily fabricated and illuminated compared with a silicon lens [10]. Figure 8 shows *S11*, gain, and efficiency with respect to frequency, where Port 1 is shown in Figure 2. Figure 9b shows that the catadioptric dielectric lens also increases antenna impedance.

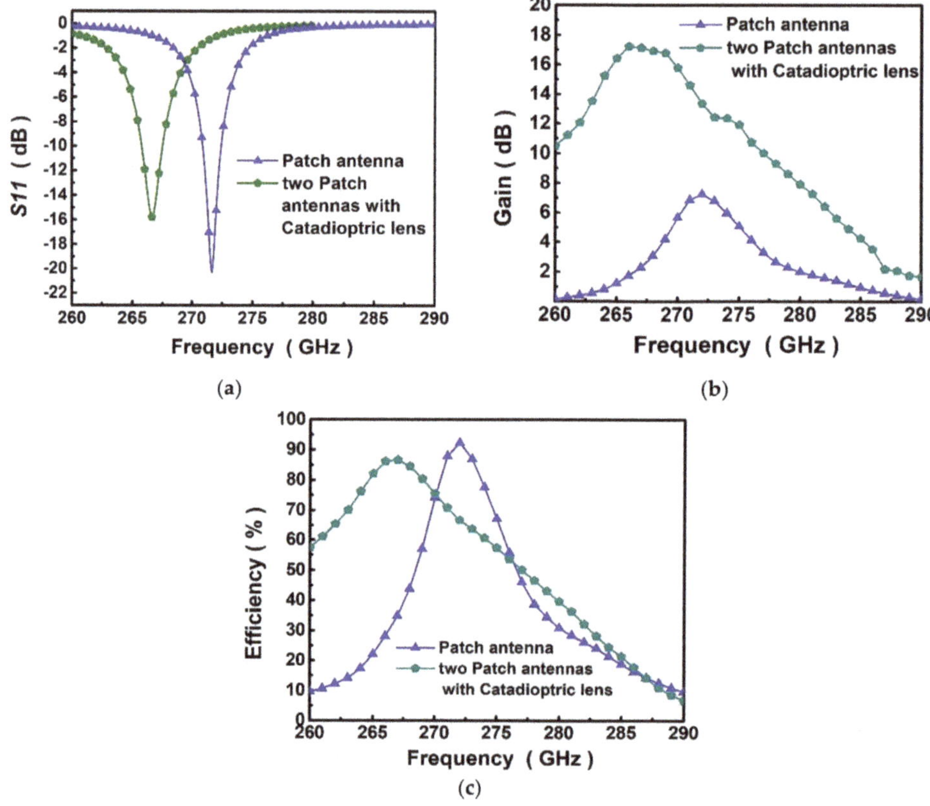

Figure 8. Simulation results for (**a**) *S11*, (**b**) gain, and (**c**) efficiency.

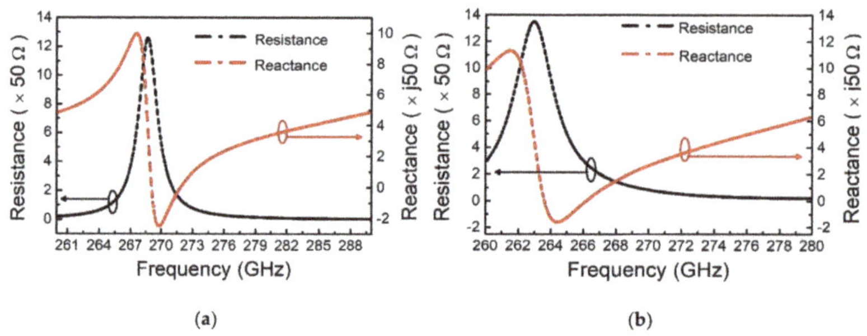

Figure 9. Output impedance: (**a**) patch antenna and (**b**) two patch antennas with catadioptric lens.

5. Terahertz Chip Measurements

The THz detector performance was modelled using Dyakonov–Shur plasma wave theory. Plasma waves at the insulator–metal interface can not only achieve THz wave detection but also achieve power amplification [11,30], where ΔU is the measured voltage from the SR830 at the amplifier output terminal, as shown in Figure 3a [31], and there is a positive correlation between ΔU and received power. Plasma waves are usually overdamped in CMOS structures [32,33], hence we propose a non-resonant broadband detector [34].

The method in [31] was used to determine voltage responsivity R_V for a detector as follows:

$$R_V = \frac{\Delta U}{P_{det}} = \frac{\int_0^L E(x)dx}{L_{THz}A_{det}} \tag{6}$$

where P_{det} is the THz power received by antennas, L is the length of the NMOSFET (TeraFET); E is the electric field intensity caused by terahertz waves, L_{THz} is the THz power density on the antennas measured with a large aperture THz power meter [31,33,35,36], as shown in Table 3, and $A_{det} \approx 0.075$ mm^2 is the rectangular inset-feed patch antenna size:

$$G_A = \frac{\Delta U}{G_T\left(\frac{\lambda}{4\pi R}\right)^2 R_V P_{BWO}} \tag{7}$$

following [21], where GT is the total gain due to the first and second lenses, and on-chip and lock-in amplifiers, as shown in Figure 10b, R = 40 cm is the distance between the backward wave oscillator (BWO) and detector, $PBWO$ is the power of the BWO wave source [21], as shown in Table 3, and λ is the wavelength. By adjusting the position of the lenses and the THz power meter, the THz power meter position of maximum L_{THz} is recorded, then the THz power meter is replaced with detector chip.

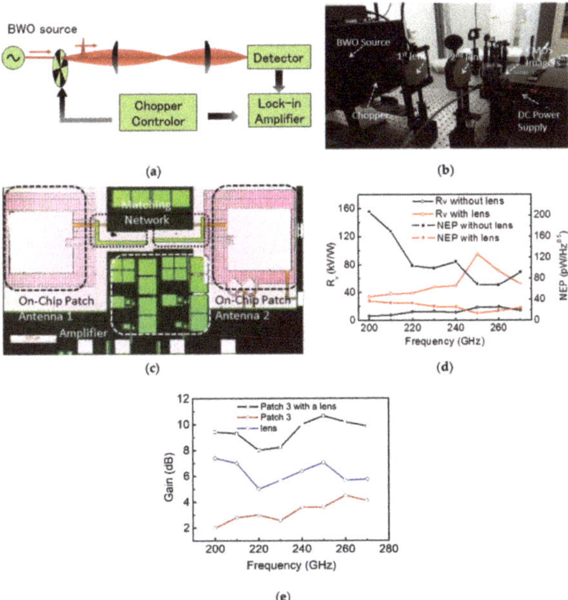

Figure 10. (**a**) Measurement setup for the detectors, (**b**) micro-photo of the chip, (**c**) detector test platform, (**d**) R_V and NEP, and (**e**) gain.

Table 3. NMOSFET test results.

Frequency (GHz)	*PBWO* (mW)	L_{THz} (mW/mm^2)	ΔU without Lens (μV)	ΔU with Lens (μV)
200	12	213	46.6	540.8
210	14	240	57.1	707.4
220	6	133	92.3	402
230	2	26	96.7	93.1
240	4	67	8.6	246.85
250	10	160	141.3	1148.04
260	16	267	144.3	1338
270	10	187	105.4	749

The gain of the horn-like lens is as follows:

$$G_L = \frac{R_{V with\ lens}}{R_{V without\ lens}} \tag{8}$$

NMOSFET channel thermal noise is the dominant noise source for a non-current biased NMOSFET detector. Noise equivalent power (NEP) determines the detector noise component with respect to the particular wavelength and measurement bandwidth, and represents the minimum optical power required for output signal-to-noise ratio = 1 [37]. Since the on-chip amplifier and detector are integrated on the same chip, low-noise amplifier noise cannot be measured separately. However, simulation shows low-noise amplifier *NEP* to be very small, hence amplifier flicker noise was negligible and need not be considered. Therefore, we estimated the detector *NEP* as follows:

$$NEP = \sqrt{4kTR_{ds}}/R_V \tag{9}$$

where k is the Boltzmann constant, T is the measurement environment temperature (nominally 300 K), and R_{ds} is half of the drain-source resistance.

The NMOSFET threshold voltage was approximately 0.45 V. The detectors were tested with gate voltages from 0.15 V to 0.36 V and bias voltage = 0.238 V for maximum R_V. The NMOSFET is most suitable to operate in the sub-threshold region, forming a two-dimensional electronic fluid, increasing channel resistance sufficiently, and most likely to obtain maximum R_V.

Figure 10e shows that the detector with catadioptric dielectric lens achieved R_V = 95.67 kV/W with NEP = 12.8 pW/Hz$^{0.5}$ at 250 GHz, whereas it achieved R_V = 19.2 kV/W with NEP = 67.2 pW/Hz$^{0.5}$ at 260 GHz without the lens. Comparing Figure 10d with Figure 8, simulated frequencies for maximum R_V differed from the measured case. The most likely cause was fabrication error. Figure 10e shows patch antenna gain with and without catadioptric lens calculated from Equations (6)–(8). G_A is relatively meaningless due to the on-chip and lock-in amplifiers, but G_L is accurate. Patch antenna gain with a catadioptric lens was slightly lower than the simulation result.

6. Conclusions

This study proposed a THz rectangular inset-feed patch antenna that can be easily achieved in CMOS technology and can be integrated perfectly with a low-noise amplifier. However, the antenna has narrow bandwidth and is sensitive to dielectric constant and shape. The dielectric constant cannot be changed by the circuit and antenna designer for standard CMOS processing, but antenna impedances and operating frequency can be controlled by reasonable on-chip antenna design considering antenna gain and bandwidth. A front-end ZEONEX RS420 catadioptric horn-like lens greatly improved voltage responsivity of the THz detector.

Two patch antennas with lens achieved 17.14 dB maximum simulated gain at 267 GHz, while the gain of two patch antennas is 4.5 dB at 260 GHz and the gain of two antennas with catadioptric lens is 10.67 dB at 250 GHz.

Author Contributions: Methodology, Investigation, Writing—original draft, and Data curation, F.Z.; Conceptualization, Supervision, and Project administration, L.M.; Resources, W.G.; Funding acquisition, S.X.; Writing—review and editing, C.A.T.H.T. All authors have read and agreed to the published version of the manuscript.

Funding: This work was supported by the National Natural Science Foundation of China (Grant No. 51425502) and the Tianjin Natural Science Foundation (Grant Nos. 18JCQNJC04800 and 18JCZDJC31800).

Acknowledgments: The authors wish to thank Cunlin Zhang, Physics Department, Capital Normal University, Beijing, China, for providing a THz BWO source to measure the detector.

Conflicts of Interest: The authors declare no conflict of interest.

References

1. Liu, Z.-Y.; Qi, F.; Wang, Y.-L.; Liu, P.-X.; Li, W.-F. A 220- to 299-GHz CMOS Terahertz Detector. *J. Infrared Millim. Terahertz Waves* **2019**, *40*, 606–619. [CrossRef]
2. Choe, W.; Jeong, J. A Broadband THz On-Chip Transition Using a Dipole Antenna with Integrated Balun. *Electronics* **2018**, *7*, 236. [CrossRef]
3. Zheng, Y.; Song, L.; Huang, J.; Zhang, H.; Fang, F. Detection of the three-dimensional trajectory of an object based on a curved bionic compound eye. *Opt. Lett.* **2019**, *44*, 4143–4146. [CrossRef] [PubMed]
4. Peng, B.; Lei, J.; Fu, H.; Zhang, C.; Chua, T.-S.; Li, X. Unsupervised Video Action Clustering via Motion-Scene Interaction Constraint. *IEEE Trans. Circuits Syst. Video Technol.* **2020**, *30*, 131–144. [CrossRef]
5. Preu, S. Introduction to the Special Issue on Terahertz Cameras and Detector Arrays. *J. Infrared Millim. Terahertz Waves* **2015**, *36*, 877–878. [CrossRef]
6. Shanawani, M.; Masotti, D.; Costanzo, A. THz Rectennas and Their Design Rules. *Electronics* **2017**, *6*, 99. [CrossRef]
7. Bauer, M.; Raemer, A.; Chevtchenko, S.A.; Osipov, K.Y.; Cibiraite, D.; Pralgauskaite, S.; Ikamas, K.; Lisauskas, A.; Heinrich, W.; Krozer, V.; et al. A High-Sensitivity AlGaN/GaN HEMT Terahertz Detector with Integrated Broadband Bow-Tie Antenna. *IEEE Trans. Terahertz Sci. Technol.* **2019**, *9*, 430–444. [CrossRef]
8. Zak, A.; Andersson, M.A.; Bauer, M.; Matukas, J.; Lisauskas, A.; Roskos, H.G.; Stake, J. Antenna-Integrated 0.6 THz FET Direct Detectors Based on CVD Graphene. *Nano Lett.* **2014**, *14*, 5834–5838. [CrossRef]
9. Qin, H.; Li, X.; Sun, J.; Zhang, Z.; Sun, Y.; Yu, Y.; Li, X.; Luo, M. Detection of incoherent terahertz light using antenna-coupled high-electron-mobility field-effect transistors. *Appl. Phys. Lett.* **2017**, *110*, 171109. [CrossRef]
10. Hadi, R.A.; Sherry, H.; Grzyb, J.; Baktash, N.; Zhao, Y.; Öjefors, E.; Kaiser, A.; Cathelin, A.; Pfeiffer, U. A broadband 0.6 to 1 THz CMOS imaging detector with an integrated lens. In Proceedings of the 2011 IEEE MTT-S International Microwave Symposium, Baltimore, MD, USA, 5–10 June 2011.
11. Zhao, F.; Zhu, C.; Guo, W.; Cong, J.; Tee, C.A.T.H.; Song, L.; Zheng, Y. Resonant Tunneling Diode (RTD) Terahertz Active Transmission Line Oscillator with Graphene-Plasma Wave and Two Graphene Antennas. *Electronics* **2019**, *8*, 1164. [CrossRef]
12. Zhao, F.; Wang, Y.; Guo, W.; Cong, J.; Tee, C.A.T.; Song, L.; Zheng, Y. Analysis of the eight parameter variation of the resonant tunneling diode (RTD) in the rapid thermal annealing process with resistance compensation effect. *AIP Adv.* **2020**, *10*, 035103. [CrossRef]
13. Ojefors, E.; Pfeiffer, U.R.; Lisauskas, A.; Roskos, H.G. A 0.65 THz Focal-Plane Array in a Quarter-Micron CMOS Process Technology. *IEEE J. Solid State Circuits* **2009**, *44*, 1968–1976. [CrossRef]
14. Han, R.; Zhang, Y.; Coquillat, D.; Hoy, J.; Videlier, H.; Knap, W.; Brown, E.; Kenneth, K.O. A 280-GHz schottky diode detector in 130-nm digital CMOS. *IEEE J. Solid State Circuits* **2011**, *46*, 2602–2612. [CrossRef]
15. Uzunkol, M.; Gurbuz, O.D.; Golcuk, F.; Rebeiz, G.M. A 0.32 THz SiGe 4x4 Imaging Array Using High-Efficiency On-Chip Antennas. *IEEE J. Solid State Circuits* **2013**, *48*, 2056–2066. [CrossRef]
16. Yang, S.Y.; Cho, C.S.; Lee, J.W.; Kim, J. Design of Sub-THz Log-Periodic Antenna for High Input Impedance. In Proceedings of the 34th International Conference on Infrared, Millimeter, and Terahertz Waves, Busan, Korea, 21–25 September 2009; pp. 739–740.
17. Wang, P.; Huang, Y.H.; Zuo, Z.G.; Yu, J.Z. Optimization of 0.34 THz log-periodic antenna for superconducting terahertz emitter. In Proceedings of the 2015 Asia-Pacific Microwave Conference (APMC), Nanjing, China, 6–9 December 2015; pp. 1–5.

18. Seok, E.; Cao, C.; Shim, D.; Arenas, D.J. A 410GHz CMOS Push-Push Oscillator with an On-Chip Patch Antenna. In Proceedings of the 2008 IEEE International Solid-State Circuits Conference, San Francisco, CA, USA, 3–7 February 2008; pp. 472–629.
19. Boppel, S.; Lisauskas, A.; Bauer, M.; Hajo, A.S.; Zdanevičius, J.; Matukas, J.; Mittendorff, M.; Winnerl, S.; Krozer, V.; Roskos, H.G. Monolithically-integrated antenna-coupled field-effect transistors for detection above 2 THz. In Proceedings of the European Conference on Antennas and Propagation, Lisbon, Portugal, 13–17 April 2015; pp. 1–3.
20. Xu, L.J.; Tong, F.C.; Xue, B.; Qin, L. Design of miniaturised on-chip slot antenna for THz detector in CMOS. *IET Microw. Antennas Propag.* **2018**, *12*, 1324–1331. [CrossRef]
21. Li, C.H.; Chiu, T.Y. 340-GHz Low-Cost and High-Gain On-Chip Higher Order Mode Dielectric Resonator Antenna for THz Applications. *IEEE Trans. Terahertz Sci. Technol.* **2017**, *7*, 284–294. [CrossRef]
22. Grzyb, J.; Zhao, Y.; Pfeiffer, U.R. A 288-GHz Lens-Integrated Balanced Triple-Push Source in a 65-nm CMOS Technology. *IEEE J. Solid State Circuits* **2013**, *48*, 1751–1761. [CrossRef]
23. Cong, J.; Mao, L.; Xie, S.; Zhao, F.; Yan, D.; Guo, W. Photoelectric Dual Control Negative Differential Resistance Device Fabricated by Standard CMOS Process. *IEEE Photonics J.* **2019**, *11*, 1–11. [CrossRef]
24. Niu, K.; Huang, Z.; Li, M.; Wu, X. Optimization of the Artificially Anisotropic Parameters in WCS-FDTD Method for Reducing Numerical Dispersion. *IEEE Trans. Antennas Propag.* **2017**, *65*, 7389–7394. [CrossRef]
25. Harrington, R.F. Time-Harmonic Electromagnetic Fields. *Siam J. Math. Anal.* **1961**, *12*, 323.
26. Schneider, M.V. Microstrip Lines for Microwave Integrated Circuits. *Bell Labs Tech. J.* **1969**, *48*, 1421–1444. [CrossRef]
27. Jameson, S.; Halpern, E.; Socher, E. A 300 GHz wirelessly locked 2*3 array radiating 5.4 dBm with 5.1% DC-to-RF efficiency in 65 nm CMOS. In Proceedings of the 2016 IEEE International Solid-State Circuits Conference, San Francisco, CA, USA, 31 January–4 February 2016; pp. 348–349.
28. Golcuk, F.; Gurbuz, O.D.; Rebeiz, G.M. A 0.39–0.44 THz 2x4 Amplifier-Quadrupler Array With Peak EIRP of 3–4 dBm. *IEEE Trans. Microw. Theory Tech.* **2013**, *61*, 4483–4491. [CrossRef]
29. Banik, B. *Photonic THz Generation and Quasioptical Integration for Imaging Applications*; Chalmers University of Technology: Goteborg, Sweden, 2009.
30. Yu, A.Q.; Guo, X.G.; Zhu, Y.M.; Balakin, A.V.; Shkurinov, A.P. Metal-graphene hybridized plasmon induced transparency in the terahertz frequencies. *Opt. Express* **2019**, *27*, 11. [CrossRef] [PubMed]
31. Schuster, F.; Coquillat, D.; Videlier, H.; Sakowicz, M.; Teppe, F.; Dussopt, L.; Giffard, B.; Skotnicki, T.; Knap, W. Broadband terahertz imaging with highly sensitive silicon CMOS detectors. *Opt. Express* **2011**, *19*, 7827–7832. [CrossRef]
32. Knap, W.; Dyakonov, M.; Coquillat, D.; Teppe, F.; Dyakonova, N.; Lusakowski, J.; Karpierz, K.; Sakowicz, M.; Valusis, G.; Seliuta, D.; et al. Field Effect Transistors for Terahertz Detection: Physics and First Imaging Applications. *J. Infrared Millim. Terahertz Waves* **2009**, *30*, 1319–1337. [CrossRef]
33. Lisauskas, A.; Roskos, H.G. Terahertz imaging with Si MOSFET focal-plane arrays. *Proc. SPIE Int. Soc. Opt. Eng.* **2009**, *7215*, 935–937. [CrossRef]
34. Dyakonov, M.; Shur, M. Shallow water analogy for a ballistic field effect transistor: New mechanism of plasma wave generation by dc current. *Phys. Rev. Lett.* **1993**, *71*, 2465. [CrossRef]
35. Lisauskas, A.; Pfeiffer, U.; Oejefors, E.; Bolivar, P.H.; Glaab, D.; Roskos, H.G. Rational design of high-responsivity detectors of terahertz radiation based on distributed self-mixing in silicon field-effect transistors. *J. Appl. Phys.* **2009**, *105*, 114511. [CrossRef]
36. Sherry, H.; Hadi, R.A.; Grzyb, J.; Öjefors, E.; Cathelin, A.; Kaiser, A.; Pfeiffer, U.R. Lens-integrated THz imaging arrays in 65 nm CMOS technologies. In Proceedings of the 2011 IEEE Radio Frequency Integrated Circuits Symposium, Baltimore, MD, USA, 5–7 June 2011; pp. 1–4.
37. Zheng, Y.; Song, L.; Hu, G.; Zhao, M.; Tian, Y.; Zhang, Z.; Fang, F. Improving environmental noise suppression for micronewton force sensing based on electrostatic by injecting air damping. *Rev. Sci. Instrum.* **2014**, *85*, 055002. [CrossRef]

Article

Terahertz Synthetic Aperture Imaging with a Light Field Imaging System

Nanfang Lyu, Jian Zuo, Yuanmeng Zhao and Cunlin Zhang *

Key Laboratory of Terahertz Optoelectronics, Ministry of Education, Beijing Key Laboratory for Terahertz Spectroscopy and Imaging, and Beijing Advanced Innovation Center for Imaging Technology, Department of Physics, Capital Normal University, 105th West 3rd Ring Road North, Beijing 100048, China; 2150602040@cnu.edu.cn (N.L.); jian.zuo@cnu.edu.cn (J.Z.); zhao.yuanmeng@cnu.edu.cn (Y.Z.)

* Correspondence: cunlin_zhang@cnu.edu.cn; Tel.: +86-10-6898-0838

Received: 29 February 2020; Accepted: 11 May 2020; Published: 18 May 2020

Abstract: In terahertz imaging systems based on Gaussian beam active illumination and focal plane array detectors, severe image distortion has been observed, which significantly reduces the resolving power of the imaging system. To solve this problem, a novel computational method, Light Field Imaging (LFI), has been introduced for terahertz imaging. A conventional transmission-type terahertz imaging system based on a gas-pumped terahertz source and terahertz Focal Plane Array Detectors (FPA) arrays is established to analyze the problem of image distortion. An experimental virtual camera array terahertz LFI system is also established. With the acquisition and reconstruction of synthetic aperture terahertz light fields, the improvement on resolving power and SNR performance have been validated.

Keywords: terahertz imaging; light field imaging; synthetic aperture imaging; image distortion; resolving power

1. Introduction

Terahertz imaging is a novel imaging method with promising application fields. Terahertz waves are electromagnetic waves with a wavelength between 30 and 3000 μm or a frequency between 0.1 and 10 THz. Terahertz waves have a low electron energy and a high penetration against dielectric materials, which make them applicable for see-through imaging for multiple materials such as plastic, ceramic, woods, papers and polymer composites [1–3]. Compared with common see-through imaging methods such as X-ray, microwave and ultrasonic imaging, terahertz imaging has advantages in penetration, non-destructive and resolving power, which make it a promising and irreplaceable imaging method in multiple application fields, including industrial Non-Destructive Testing (NDT)—[4,5], heritage conservation [6–8], aerospace [9], security [10–12], biology and medicine [13].

Thanks to the fast development of terahertz source and detector component techniques, terahertz Focal Plane Array Detectors (FPA) based on a microbolometer [14,15] and complementary metal oxide semiconductor (CMOS) [16–19] make fast, large-scale, high-resolution, incoherent terahertz imaging methods available. However, passive mode imaging in the terahertz region is infeasible for the see-through applications mentioned above, as little background radiation in the terahertz region exists in the environment, and the method is still limited by the performance of commercially available devices [20]. To ensure an acceptable signal-to-noise ratio (SNR) and resolve the power for samples with a certain thickness, most terahertz imaging applications rely on a high-power, collimated source for illumination [3]. The beam of a high power continuous-wave (CW) terahertz source such as a gas-pumped laser and quantum cascade laser (QCL) is a Gaussian beam. When used in active illumination imaging, the wavefront characteristics of the Gaussian beam would lead to distortion, which significantly reduces the resolving power. A beam homogenizer would help in reducing such distortion; however, a beam homogenizer diffuses beam energy, which leads to negative impacts in transmission-type imaging applications that require a sufficiently effective imaging depth and concentrated beam energy. Moreover, the microbolometer and CMOS terahertz FPA have issues in terms of Fixed Pattern Noise (FPN) [21,22], which leads to adverse effects on the SNR and the resolving power of the imaging result.

An optical system with a larger aperture will help to solve the problems mentioned above. However, limited by the size of the system and the cost of processing, enlarging the aperture size is impracticable in most applications. Introducing the Light Field Imaging (LFI) method will help to solve this problem. LFI is an incoherent computational imaging method which extracts information from data of higher dimensions and achieves a better performance compared with conventional methods [23]. In LFI, both the positional and directional information of the imaging object are gathered simultaneously, implementing enhancements in SNR, dynamic range, depth of field (DoF) and additional depth information about the images [24] with reconstruction algorithms. In the visible region, light field acquisition methods based on microlens arrays [24], camera arrays [25], and coded aperture compressive sensing [26] are now being widely studied. Moreover, the LFI technique is also utilized in multiple applications including dynamic refocusing [24], DoF extension [27], all-in-focus reconstruction [28], depth estimation [29,30], 3D reconstruction [31], super-resolution reconstruction [32,33], synthetic aperture imaging and high-speed photography [27].

Jain et al. [34] proved the feasibility of LFI in the terahertz waveband with a simplified experimental setup. In their experiments, the LFI of a point source and simple opaque object was implemented with a camera module of a low angular (approximately 1.5°) and spatial (approximately 28 mm) resolution. In their study, the effects caused by the coherence of light sources and the aperture of the imaging system are not further discussed, which are important factors for practical imaging systems and objects. In fact, these effects cause the main differences in the image characteristics between the terahertz and visible region.

In this paper, a novel, incoherent synthetic aperture terahertz imaging method based on the LFI technique is demonstrated, and the performance improvements of the new method are validated. The phenomenon of image distortion happening in the terahertz imaging system with Gaussian beam illumination has been investigated. With microbolometer-based terahertz FPA detectors and a high-power gas-pumped terahertz source, a terahertz virtual camera array LFI system has been established. With the experiments of transmission-type imaging configuration, the effectiveness of the LFI method in removing the image distortion and improving the resolving power compared with a conventional setup is validated.

2. Theory and Methods

2.1. Light Field Theory and Mathematical Model

The plenoptic model is a model used to describe the "light field", i.e., the spatial propagation of light flowing. In particular, in the plenoptic model, the light flow is simplified as sizeless "rays", described by their position and direction and by the plenoptic function. Assuming that the rays emitted by light sources do not spatially overlap, the positional and directional information of light rays can be described using four dimensions.

In the case of a two-plane scheme, a 4D light field can be described with a 4D plenoptic function $L_F(x, y, u, v)$, which describes rays with the intersectional coordinates of two parallel planes.

Specifically, two planes uv and xy are defined with a fixed distance, and the rays will intersect with two lanes. The combination of all the rays passing through every coordinate on planes uv and xy create the full light field, as shown in Figure 1. The two-plane scheme is introduced for practical light field acquisition methods. For any combination of two dimensions in the 4D plenoptic function, for example, rays intersect with a certain (u, v) or certain (x, y) coordinate—we call this a 2D slice of the

4D plenoptic function. In practice, the "slice" is used to describe the sub-aperture images, macro-pixels or epipolar images of 4D light fields.

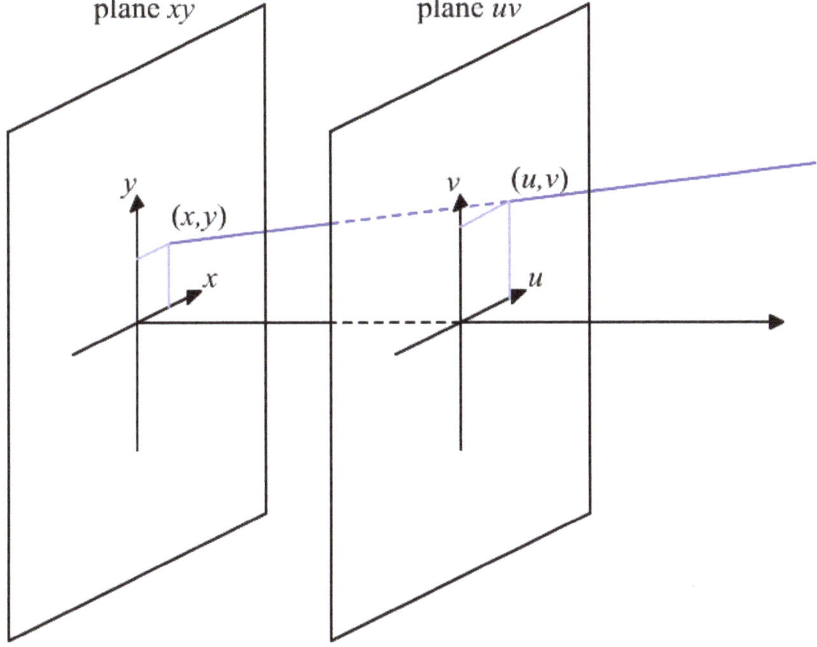

Figure 1. The two-plane scheme of 4D light field.

The procedure of LFI can be divided into two steps: the acquisition and the reconstruction. The acquisition phase physically gathers and resolves the 4D light field of the object, and the reconstruction phase generates enhanced images with the acquired 4D light field information.

2.2. The Acquisition of the 4D Light Field

Compared with conventional imaging, the acquisition of the 4D light field initiates both positional and directional resolution simultaneously, namely the resolving of individual light rays. Imagine an FPA imaging system, or "camera": light rays are focused by optic systems (called the main aperture) at the uv plane, then gathered, resolved and imaged by detector pixels at the xy plane. The pixels are the partitions of xy planes. The light signal received by pixels corresponds to the sum of light rays emitted from a certain position in all directions (which can be gathered by the main aperture). The light signal received by pixels is converted into images. We perform the same partition on the uv plane, and the partition on the uv plane is called a "sub-aperture". The lightbeams illuminated on a single pixel through the single sub-aperture could be approximately regarded as rays in the plenoptic model. Rays correspond to the partition of light emitted by a certain point source along a certain direction. In this case, besides the image plane in the image space, the conjugated plane of the xy plane in the object space can be regarded as the xy plane as well, after a simple coordinate transformation, as shown in Figure 2a.

For an FPA imaging system, when the imaging distance is long enough, relative to the focal length, the size of the aperture can be ignored, and the whole beam focused on a single pixel through the aperture can be regarded as an individual ray. Thus, the image recorded by the detector array can be regarded as a combination of rays in certain positions and different directions, namely a 2D slice of the 4D light field. Arranging such cameras on the *uv* plane, the full light field can be recorded by recording all of its sub-aperture slices. With this camera array, the 4D light field can be acquired with a synthetic aperture equivalent to the size of the camera array, and its resolution depends on the resolution and spacing of the camera modules. In this case, the *xy* plane does not really exist in the image space, since there is no common image space for every single camera. Besides this, the common plane conjugated to image planes of all cameras (supposing they are regularly arranged and have the same optical parameters) in the object space is regarded as the *xy* plane, as shown in Figure 2b. In practice, the light field slices are acquired with a camera array consisting of multiple FPA cameras, or a virtual camera array for static light fields, with a limited number of cameras.

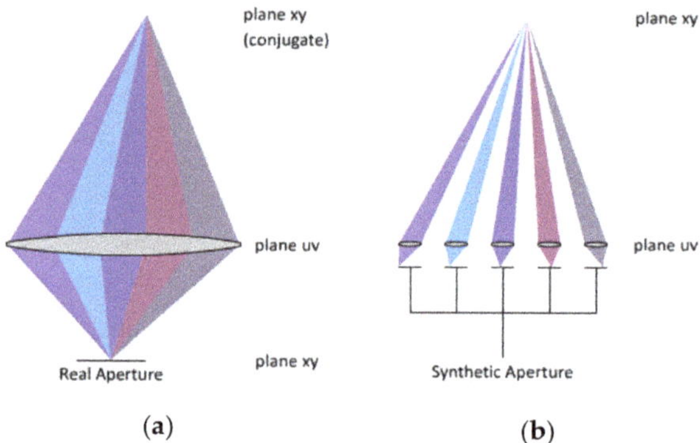

Figure 2. Light field acquisition and partition in (**a**) real aperture and (**b**) synthetic aperture.

2.3. Light Field Reconstruction

The reconstruction of a 4D light field is a simulation of the focusing and imaging procedures of the optical imaging system in a geometric optic manner, based on the 4D light field information.

For a given 4D light field $L_F(x,y,u,v)$, the planes, *uv*, *xy*, and the distance *F* between planes *uv* and *xy* are given.

When replacing the plane *xy* with the plane $x'y'$ in the description, where the distance from plane *uv* is $z = \alpha F$, the transformed plenoptic function could be described as

$$L_z(x',y',u,v) = L_F\left(u\left(1-\frac{F}{z}\right)+x'\frac{F}{z}, v\left(1-\frac{F}{z}\right)+y'\frac{F}{z}, u, v\right) \tag{1}$$

which is equal to shearing the plenoptic function in *xy* dimensions. The ray passing through the point (x_0, y_0, z) in space and the point $(u,v,0)$ on the plane *uv* could be described as

$$S_{x_0,y_0,z}(u,v) = L_F\left(u\left(1-\frac{F}{z}\right)+x_0\frac{F}{z}, v\left(1-\frac{F}{z}\right)+y_0\frac{F}{z}, u, v\right) \tag{2}$$

Substituting the deferent points (u,v) in Equation (2) gives us the combination of the rays passing through the point (x_0, y_0, z). Assuming we know the depth $z(x_0, y_0)$ of all points (x_0, y_0). in the scene, the Equation (2) could be transformed into

$$\begin{aligned} S_{x_0,y_0,z(x_0,y_0)}(u,v) &= L_F\left(u\left(1-\frac{F}{z(x_0,y_0)}\right)+x_0\frac{F}{z(x_0,y_0)}, v\left(1-\frac{F}{z(x_0,y_0)}\right)\right. \\ &\left.+y_0\frac{F}{z(x_0,y_0)}, u, v\right) \end{aligned} \tag{3}$$

and the reconstructed image could be described as

$$\begin{aligned} E(x_0, y_0) &= \iint L_{z(x_0,y_0)}(x', y', u, v)\mathrm{d}u\mathrm{d}v \\ &= \iint L_F\left(u\left(1 - \frac{F}{z(x_0,y_0)}\right) + x_0\frac{F}{z(x_0,y_0)}, v\left(1 - \frac{F}{z(x_0,y_0)}\right) + y_0\frac{F}{z(x_0,y_0)}, u, v\right)\mathrm{d}u\mathrm{d}v \end{aligned} \quad (4)$$

As mentioned in Section 2.2, in practical light field acquisition, the 4D light field is discretized into pixels in all $xyuv$ dimensions, and the slices of the light field in uv and xy dimensions are discretized into the 2D combination of pixels as well. Therefore, in the practical implementation of the reconstruction algorithm, the continuous integration in Equation (3) is transformed into the numerical sum of discretized pixels. For every discretized pixel (x_0,y_0) in the reconstructed image, the algorithm traverses all the slices (u,v) of the light field, searches for the pixels corresponding to the ray through (u,v), (x',y') and (x_0,y_0), and calculates the pixel values from all the slices.

If the input light field is from a single physical aperture, the algorithm simply simulates the imaging procedure in geometric optics and generates reconstructed images as a result. If the input light field is from the camera array described in Section 2.2, the algorithm generates reconstructed images based on a synthetic aperture.

3. Experiment and Result

3.1. Distortion Analysis of Gaussian Beam Active Illuminated Terahertz Imaging

In this section, the phenomenon of the image distortion observed in the active illuminated terahertz imaging process will be discussed.

In our experiment, a transmission-type terahertz imaging system based on Gaussian beam active illumination and an FPA terahertz camera has been established. In particular, the terahertz beam from the source is collimated by a Teflon collimation lens, transmitted through the object, and then received by the camera at different positions. In the experiment, the object and the source are fixed. The camera moves around and acquires images at different positions. The camera remains static when acquiring images to avoid motion blur. The experimental setup is shown in Figure 3.

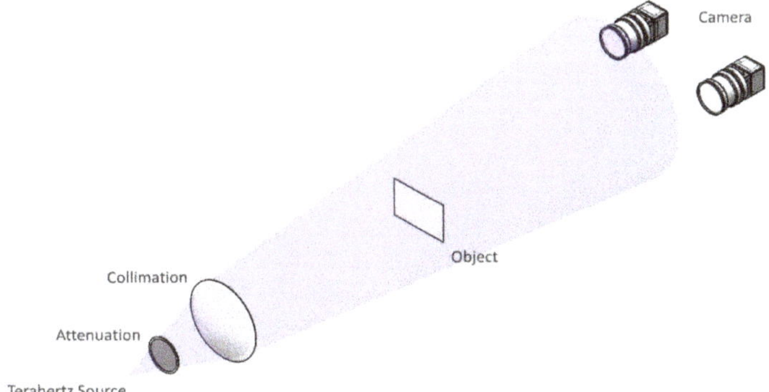

Figure 3. Experimental setup of transmission-type terahertz imaging system.

In the experiment, a Coherent SIFIR–50 gas-pumped continuous-wave terahertz source is used as an illumination source. The terahertz source works at 2.52 THz, and the output power is about 50 mW, which provides a stable and good SNR performance. An INO IRXCAM-THz–384 terahertz camera module is used to acquire terahertz images. The camera uses FPA detectors based on uncooled microbolometers, with a 35 µm pixel size and a resolution of 384 × 288 pixels, and is capable of capturing images at a 50 fps rate. The camera module has Silicon imaging optics, with a 44-mm focal length, a field of view of about 17.3 × 13.0 degrees, and a spatial resolution of about 0.5 mm. In the experiment, an extra Teflon filter is mounted in front of the camera module to filter the unwanted infrared signal in the environment.

The imaging object of the experiment is a polypropylene composite Pelican case with bumped markings. The thickness of the case is about 4 mm, and the diameter and thickness of the bumped markings are about 5 mm and 0.2 mm, respectively, as shown in Figure 4. In the experiment, the distance between the source and the object is 2000 mm, and the distance between the object and the camera is 600 mm.

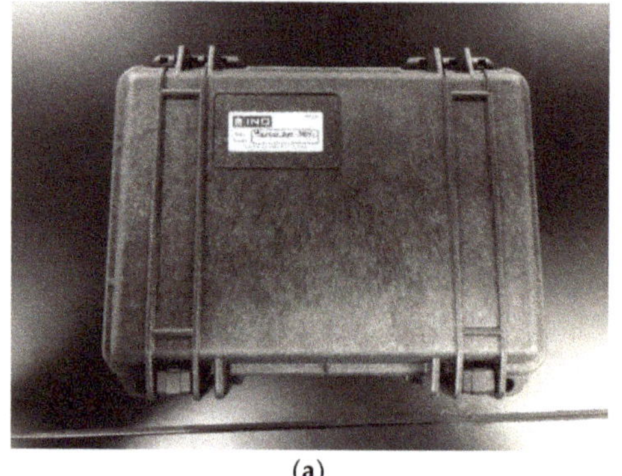

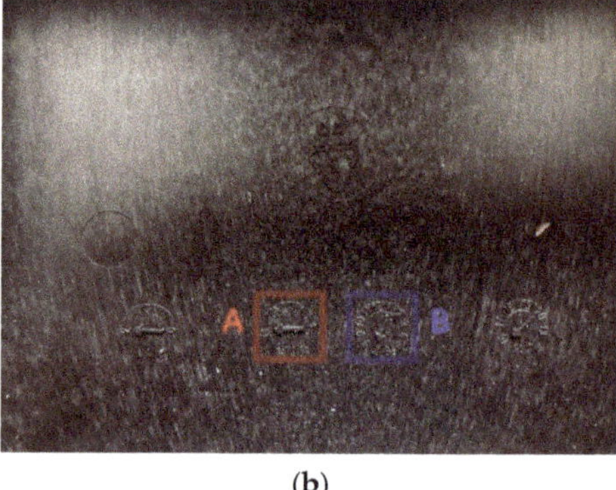

(a) (b)

Figure 4. Visual photograph of (**a**) the polypropylene composite Pelican case and (**b**) the bumped markings, A and B.

Figure 5 is the image acquired by the camera when in different positions. For active illuminated imaging, when using a point source or collimated beam for illumination, only a portion of the beam that passes through both the object and the aperture contributes to the imaging result. This portion of the "effective" beam corresponds to a "bright spot" or "bright circle" in the terahertz image. Refraction and total reflection occur at the edges of the bumped markings, which results in a significant contrast in the acquired terahertz images.

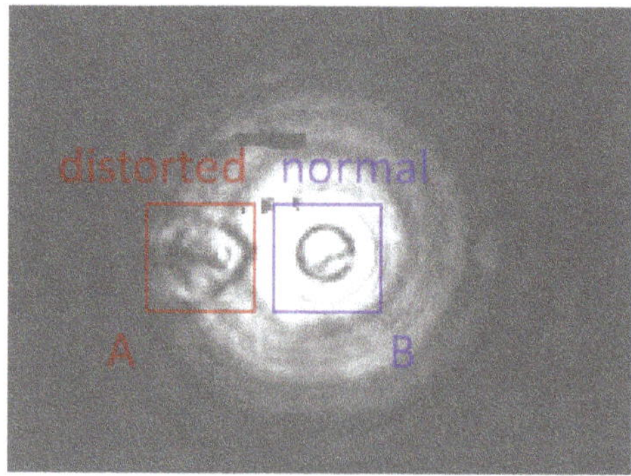

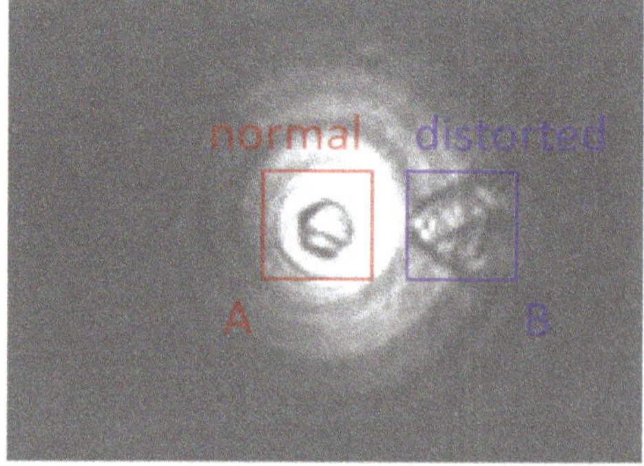

Figure 5. Terahertz images of the bumped markings, A and B, with the markings at the center and the edge of the "effective" light spot, captured at different positions, with different markings at the center of the spot.

Ideally, the round bumped markings should be the same shape in the terahertz image as their original shapes. In the experiment, when located in the center of the spot, the markings are imaged correctly as a circle. However, when located near the edge of the spot, the markings are significantly distorted and appear to be drop-shaped. The phenomenon of distortion does not occur when illuminated by a planar homogenized source or incoherent sources in the infrared waveband. In practical applications, this distortion will significantly affect the resolving power of the imaging.

3.2. Resolving Power Analysis of Terahertz LFI Setup

In this section, compared with conventional transmission-type imaging, the improvement of terahertz LFI on the resolving power and image distortion mentioned in Section 3.1 will be discussed.

Based on the transmission-type terahertz imaging system mentioned in Section 3.1, a terahertz LFI system has been established. A virtual camera array based on a terahertz camera module and motorized translation stages are utilized to replace the single terahertz camera, in order to acquire the full 4D light field.

The camera module is fixed on a 2D translation stage, which can move in both x and y directions vertical to the optical axis of the camera. The translation stages drive the camera and acquire the slices of the full light field in different positions.

In the experiment, the Coherent SIFIR–50 source works at the frequency of 1.40 THz, with an output power of about 84 mW, which provides a better penetration for the sample used in this part of the experiment. The experimental setup is shown in Figure 6.

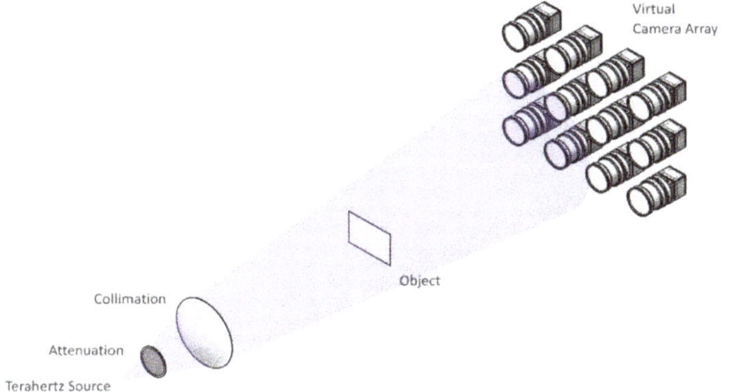

Figure 6. Experimental setup of terahertz virtual camera array system.

The imaging object is a resolving power test target with concentric circle and line pair patterns. The base material of the card is epoxy of 0.5 mm thickness, which is transparent to the terahertz wave. The concentric circle and line pair patterns are printed on the base with copper, which is opaque to the terahertz wave. The spacing between line pairs ranges from 0.125 mm to 3 mm, and the spacing between concentric circles ranges from 0.5 mm to 3 mm. The line widths of both line pairs and concentric circles are 1 mm, as shown in Figure 7. In the light field acquisition, the number of sub images is 11 × 31 sub-images, and the spacing is 5 × 5 mm. The distance between the source and the object is 2000 mm, and the distance between the object and the camera is 600 mm.

Figure 8 is one of the enlarged sub-aperture images; i.e., the imaging result of a single shot with a single camera. Compared with the results in Section 3.1, the result of line pairs and concentric circles suffered a severe and irregular distortion, which significantly reduced the resolving power of the imaging system for the patterns and made the positions and spacing of lines and circles almost indistinguishable. Moreover, the interference fringes and fixed pattern noise of the detectors further reduced the performance of the imaging result.

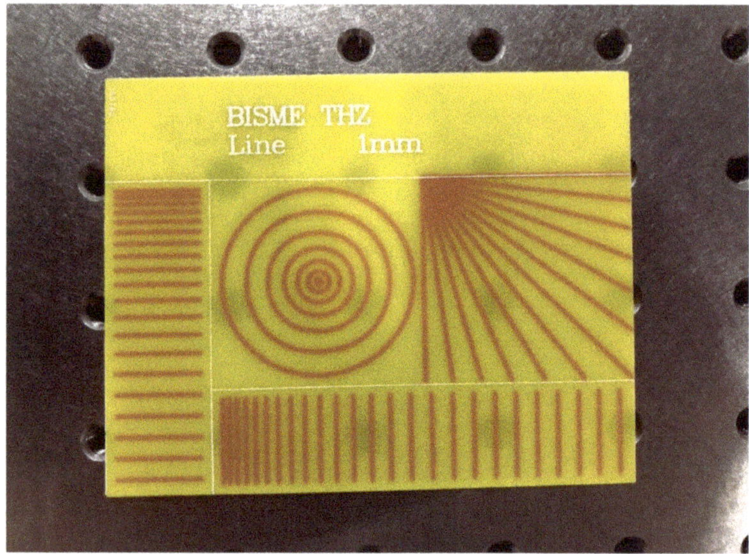

Figure 7. Visual photograph of the resolving power test target.

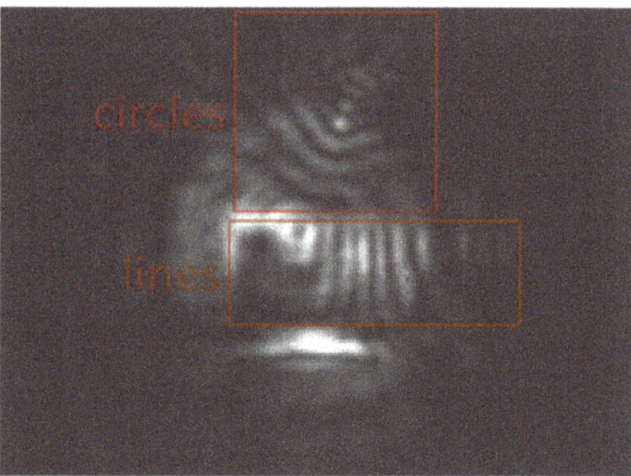

Figure 8. Terahertz image of the test target, captured by single camera in single shot.

Figure 9a,b shows the results of the light field reconstruction. Figure 9a is the result directly reconstructed by the 4D light field reconstruction algorithm, and Figure 9b is the result further enhanced with the Unsharp Mask (USM) algorithm. In reconstructed light field images, the complete portions of the object are clearly imaged. The distortion is almost removed, and the line and circle patterns are clearly distinguishable. The interference fringes and FPN are significantly suppressed as well. In the directly reconstructed image, the fourth line and the first circle with a spacing of 0.5 mm are clearly distinguishable. In the USM-enhanced image, although the suppression of noise is weakened, the resolving power of patterns is increased, since the third line, with a spacing of 0.375 mm, becomes distinguishable. Compared with the result of the conventional method, the performance and resolving power are significantly improved by using the synthetic aperture method, and the unrecognizable patterns as a result of the conventional method become fully recognizable with the synthetic aperture method.

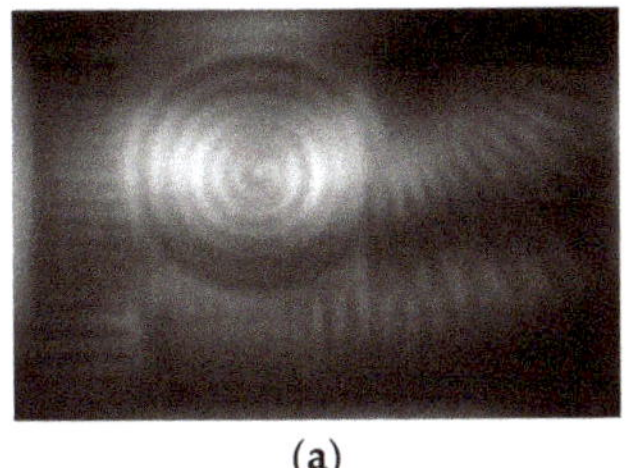

(a)

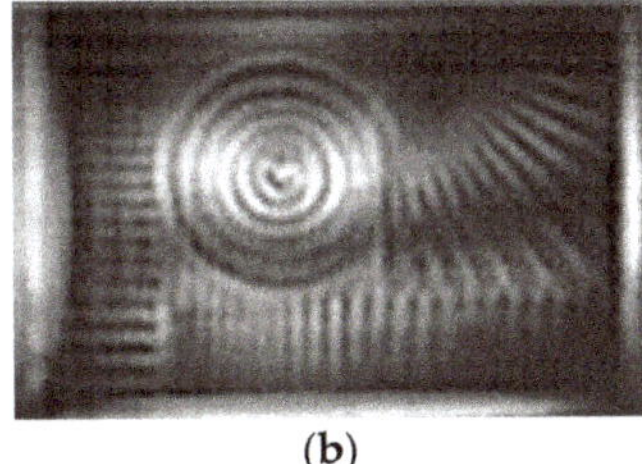

(b)

Figure 9. Reconstructed image form full light field data, (**a**) directly reconstructed and (**b**) enhanced with Unsharp Mask (USM) algorithm.

4. Conclusions

In this paper, an experimental transmission-type terahertz imaging system and a virtual camera array terahertz LFI system were established based on a high-power gas-pumped terahertz source and terahertz focal plane array detectors. On the one hand, the phenomenon of image distortion occurred in terahertz imaging, and its influences on resolving power have been observed and analyzed; on the other hand, a novel synthetic aperture method based on the LFI technique was introduced, and its improvement of image performance and suppression of image distortion were validated in the experiments carried out.

In the imaging experiment on a resolving power test target, the conventional method showed an unacceptable result, where the distortion made the line and circle patterns completely indistinguishable, while the synthetic aperture method with USM sharpening enhancement showed a much better result: the line pairs with less than 0.5 mm spacing were clearly resolved.

The application field of terahertz imaging requires imaging methods with high efficiency, high performance and compact integration. Computational imaging methods such as LFI are promising approaches to overcome the disadvantages of current imaging methods caused by the limitations in terahertz sources and detector technologies and to fulfil further requirements in other potential application fields.

Author Contributions: Conceptualization, C.Z., N.L., J.Z., and Y.Z.; methodology, N.L.; software, N.L.; validation, C.Z., N.L., J.Z. and Y.Z.; formal analysis, N.L.; investigation, N.L.; resources, C.Z., J.Z. and Y.Z.; writing—original draft preparation, N.L.; writing—review and editing, C.Z.; visualization, N.L.; supervision, C.Z.; project administration, C.Z.; funding acquisition, C.Z. All authors have read and agreed to the published version of the manuscript.

Funding: This research was funded by the National Key Scientific Instrument and Equipment Development Projects of China, grant number DH–2012YQ14005; and National Basic Research Program of China (973 Program), grant number 2014CB339806.

Conflicts of Interest: The authors declare no conflict of interest.

References

1. Mittleman, D.M.; Jacobsen, R.; Nuss, M. T-ray imaging. *IEEE J. Sel. Top. Quantum Electron.* **1996**, 2, 679–692. [CrossRef]
2. Guillet, J.-P.; Recur, B.; Frederique, L.; Bousquet, B.; Canioni, L.; Manek-Hönninger, I.; Desbarats, P.; Mounaix, P. Review of Terahertz Tomography Techniques. *J. Infrared Millim. Terahertz Waves* **2014**, 35, 382–411. [CrossRef]
3. Mittleman, D.M. Twenty years of terahertz imaging [Invited]. *Opt. Express* **2018**, 26, 9417–9431. [CrossRef] [PubMed]
4. Zhong, H.; Xu, J.; Xie, X.; Yuan, T.; Reightler, R.; Madaras, E.; Zhang, X. Nondestructive defect identification with terahertz time-of-flight tomography. *IEEE Sens. J.* **2005**, 5, 203–208. [CrossRef]

5. Tanabe, T.; Watanabe, K.; Oyama, Y.; Seo, K. Polarization sensitive THz absorption spectroscopy for the evaluation of uniaxially deformed ultra-high molecular weight polyethylene. *NDT E Int.* **2010**, *43*, 329–333. [CrossRef]
6. Groves, R.M.; Pradarutti, B.; Kouloumpi, E.; Osten, W.; Notni, G. 2D and 3D non-destructive evaluation of a wooden panel painting using shearography and terahertz imaging. *NDT E Int.* **2009**, *42*, 543–549. [CrossRef]
7. Jackson, J.B.; Mourou, M.; Whitaker, J.; Duling, I.; Williamson, S.; Menu, M.; Mourou, G.; Iii, I.D. Terahertz imaging for non-destructive evaluation of mural paintings. *Opt. Commun.* **2008**, *281*, 527–532. [CrossRef]
8. Caumes, J.-P.; Younus, A.; Salort, S.; Chassagne, B.; Recur, B.; Ziéglé, A.; Dautant, A.; Abraham, E. Terahertz tomographic imaging of XVIIIth Dynasty Egyptian sealed pottery. *Appl. Opt.* **2011**, *50*, 3604. [CrossRef]
9. Stoik, C.; Bohn, M.; Blackshire, J. Nondestructive evaluation of aircraft composites using reflective terahertz time domain spectroscopy. *NDT E Int.* **2010**, *43*, 106–115. [CrossRef]
10. Kapilevich, B.; Pinhasi, Y.; Arusi, R.; Anisimov, M.; Hardon, D.; Litvak, B.; Wool, Y. 330 GHz FMCW Image Sensor for Homeland Security Applications. *J. Infrared Millim. Terahertz Waves* **2010**, *31*, 1370–1381. [CrossRef]
11. Marchese, L.E.; Terroux, M.; Dufour, D.; Bolduc, M.; Chevalier, C.; Généreux, F.; Jerominek, H.; Bergeron, A. Case study of concealed weapons detection at stand-off distances using a compact, large field-of-view THz camera. In *Micro-and Nanotechnology Sensors, Systems, and Applications VI*; SPIE: Baltimore, MD, USA, 2014. [CrossRef]
12. Kemp, M.C.; Taday, P.F.; Cole, B.E.; Cluff, J.A.; Fitzgerald, A.J.; Tribe, W.R. *Security Applications of Terahertz Technology*; SPIE: Orlando, FL, USA, 2003; pp. 44–52. [CrossRef]
13. Humphreys, K.; Loughran, J.; Grądziel, M.; Lanigan, W.; Ward, T.; Murphy, J.; O'Sullivan, C. Medical applications of terahertz imaging: A review of current technology and potential applications in biomedical engineering. In Proceedings of the the 26th Annual International Conference of the IEEE Engineering in Medicine and Biology Society, San Francisco, CA, USA, 1–5 September 2005; Volume 3, pp. 1302–1305.
14. Marchese, L.; Bolduc, M.; Tremblay, B.; Doucet, M.; Oulachgar, H.; Le Noc, L.; Williamson, F.; Alain, C.; Jerominek, H.; Bergeron, A. A microbolometer-based THz imager. In Proceedings of the SPIE Defense, Security, and Sensing, Orlando, FL, USA, 5–9 April 2010. [CrossRef]
15. Dufour, D.; Marchese, L.; Terroux, M.; Oulachgar, H.; Généreux, F.; Doucet, M.; Mercier, L.; Tremblay, B.; Alain, C.; Beaupré, P.; et al. Review of terahertz technology development at INO. *J. Infrared Millim. Terahertz Waves* **2015**, *36*, 922–946. [CrossRef]
16. Al Hadi, R.; Sherry, H.; Grzyb, J.; Zhao, Y.; Forster, W.; Keller, H.M.; Cathelin, A.; Kaiser, A.; Pfeiffer, U.R. A 1 k-Pixel Video Camera for 0.7–1.1 Terahertz Imaging Applications in 65-nm CMOS. *IEEE J. Solid-State Circuits* **2012**, *47*, 2999–3012. [CrossRef]
17. Han, R.; Zhang, Y.; Kim, Y.; Shichijo, H.; Afshari, E.; O, K.K. Active Terahertz Imaging Using Schottky Diodes in CMOS: Array and 860-GHz Pixel. *IEEE J. Solid-state Circuits* **2013**, *48*, 2296–2308. [CrossRef]
18. Al Hadi, R.; Grzyb, J.; Heinemann, B.; Pfeiffer, U.R. A Terahertz Detector Array in a SiGe HBT Technology. *IEEE J. Solid-State Circuits* **2013**, *48*, 2002–2010. [CrossRef]
19. Corcos, D.; Kaminski, N.; Shumaker, E.; Markish, O.; Elad, D.; Morf, T.; Drechsler, U.; Saha, W.T.S.; Kull, L.; Wood, K.; et al. Antenna-Coupled MOSFET Bolometers for Uncooled THz Sensing. *IEEE Trans. Terahertz Sci. Technol.* **2015**, *5*, 902–913. [CrossRef]
20. Yeom, S.; Lee, D.-S.; Son, J.-Y.; Jung, M.-K.; Jang, Y.; Jung, S.-W.; Lee, S.-J. Real-time outdoor concealed-object detection with passive millimeter wave imaging. *Opt. Express* **2011**, *19*, 2530–2536. [CrossRef]
21. Coelho, P.; Tapia, J.E.; Pérez, F.; Torres, S.N.; Saavedra, C. Infrared light field imaging system free of fixed-pattern noise. *Sci. Rep.* **2017**, *7*, 13040. [CrossRef]
22. Chen, C.; Yi, X.; Zhao, X.; Xiong, B. Characterizations of VO2-based uncooled microbolometer linear array. *Sens. Actuators A Phys.* **2001**, *90*, 212–214. [CrossRef]
23. Levoy, M. Light Fields and Computational Imaging. *Computer* **2006**, *39*, 46–55. [CrossRef]
24. Ng, R. Digital Light Field Photography. Ph.D. Thesis, Stanford University, Stanford, CA, USA, 2006.
25. Wiburn, B. High Performance Imaging using Arrays of Inexpensive Cameras. Ph.D. Thesis, Stanford University, Stanford, CA, USA, 2004.
26. Liu, L.; Zhang, Z.; Gan, L.; Shen, Y.; Huang, Y. Terahertz imaging with compressed sensing. In Proceedings of the 2016 IEEE 9th UK-Europe-China Workshop on Millimetre Waves and Terahertz Technologies (UCMMT), Qingdao, China, 5–7 September 2016; pp. 50–53. [CrossRef]

27. Georgiev, T.; Lumsdaine, A. *Depth of Field in Plenoptic Cameras*; Eurographics (Short Papers): Munich, Germany, 2009.
28. Takahashi, K.; Kubota, A.; Naemura, T. Focus measurement and all in-focus image synthesis for light-field rendering. *Syst. Comput. Jpn.* **2005**, *37*, 1–12. [CrossRef]
29. Adelson, E.; Wang, J. Single lens stereo with a plenoptic camera. *IEEE Trans. Pattern Anal. Mach. Intell.* **1992**, *14*, 99–106. [CrossRef]
30. Zhang, Y.; Lv, H.; Liu, Y.; Wang, H.; Wang, X.; Huang, Q.; Xiang, X.; Dai, Q. Light-Field Depth Estimation via Epipolar Plane Image Analysis and Locally Linear Embedding. *IEEE Trans. Circuits Syst. Video Technol.* **2017**, *27*, 739–747. [CrossRef]
31. Lüke, J.P.; Marichal-Hernández, J.G.; Rosa, F.; Rodríguez-Ramos, J.M. A prototype of a real-time single lens 3D camera. Proceedings of International Conference on 3D Systems and Applications, Tokyo, Japan, 19–21 May 2010; pp. 19–106.
32. Bishop, T.E.; Zanetti, S.; Favaro, P. Light field superresolution. In Proceedings of the 2009 IEEE International Conference on Computational Photography (ICCP), San Francisco, CA, USA, 16–17 April 2009; pp. 1–9.
33. Carles, G.; Downing, J.; Harvey, A.R. Super-resolution imaging using a camera array. *Opt. Lett.* **2014**, *39*, 1889–1892. [CrossRef] [PubMed]
34. Jain, R.; Grzyb, J.; Pfeiffer, U.R. Terahertz Light-Field Imaging. *IEEE Trans. Terahertz Sci. Technol.* **2016**, *5*, 1–9. [CrossRef]

Article

Application of Terahertz Spectroscopy to Rubber Products: Evaluation of Vulcanization and Silica Macro Dispersion

Yasuyuki Hirakawa [1,*], Yuki Yasumoto [1], Toyohiko Gondo [1], Ryota Sone [2], Toshiaki Morichika [2], Takakazu Minato [2] and Masahiro Hojo [2]

[1] National Institute of Technology (KOSEN), Kurume College, Kurume 830-8555, Japan; yasugr4@gmail.com (Y.Y.); gonmota.9411@gmail.com (T.G.)

[2] Advanced Materials Division, Bridgestone Corporation, Kodaira 187-8531, Japan; ryota.sone@bridgestone.com (R.S.); toshiaki.morichika1@bridgestone.com (T.M.); takakazu.minato@bridgestone.com (T.M.); masahiro.hojo@bridgestone.com (M.H.)

* Correspondence: hirakawa@kurume-nct.ac.jp; Tel.: +81-942-35-9381

Received: 27 March 2020; Accepted: 18 April 2020; Published: 20 April 2020

Abstract: Industrial applications of terahertz (THz) technology are becoming more widespread. In particular, novel evaluation methods for essential rubber products are being developed. THz absorbance spectra of various rubber polymers and reagents enable visualization of filler dispersions and vulcanization reactions. Here, improved visualization of the vulcanization reaction in thick rubber samples is discussed. Silica macro-dispersion is also analyzed because it is a general filler in automobile tires and has been difficult to evaluate with conventional techniques.

Keywords: THz-TDS; rubber; vulcanization; silica dispersion

1. Introduction

Terahertz (THz) waves are located in the electromagnetic band between the far-infrared and the microwave regions corresponding to the fingerprint region of some materials and can exhibit penetration of materials except for metals [1]. Many applications using THz light have been attempted in various fields such as semiconductors, biological materials, and polymers [2–6]. Among them, industrial applications are limited so far, and a few, such as characterization of paint films [7] and tablet coatings [8], were reported.

Rubber products, such as automobile tires, rubber bands, rubber balls, insulators, and O-rings, are essential to modern society. Manufacturers of these products use sophisticated chemical reagents and control systems; however, the manufacturing processes depend considerably on empirical procedures and craftsmanship. This situation suggests that the production is not very efficient and could be improved. Hence, to obtain better quality of rubber products, it is crucial to realize uniform dispersions of filler materials and to visualize vulcanization.

Our research group has been developing novel evaluation methods using THz radiation. THz absorbance spectra of various chemical materials used in rubber products has been reported [9,10], and visualization of the dispersion of carbon black (CB), a major filler material, has been experimentally demonstrated [10]. It has also been reported that THz light could nondestructively detect the mesh network structure in the vulcanization reaction and the depth of the cure condition in rubber products having thicknesses greater than 10 mm [10]. These techniques are based on the transmission capability of THz radiation.

In this article, improved imaging of vulcanization depths in thick rubber samples and, for the first time, visualization of silica filler dispersion are demonstrated. The latter has been difficult to perform with conventional techniques.

2. Experimental

THz time-domain spectroscopy (THz-TDS) was performed with a custom instrument (Otsuka Electronics Co., Ltd., TR-100KS) that included a femtosecond (fs) fiber laser (IMRA AMERICA, INC., Femtolite, 100 fs, 1620 nm) and a 4′-dimethylamino-N-methyl-4-stilbazolium tosylate (DAST) crystal for THz generation. Due to the THz detector of a photoconductive switch, the bandwidth of the applicable THz radiation was limited to 3 THz for a better signal–noise ratio. The THz-TDS system was described in detail previously [10]; the transmission mode was used in this study.

The rubber samples were based on the synthetic polymer styrene-butadiene rubber (SBR, commercial-grade), which is widely used in automobile tires, and isoprene rubber (IR, commercial-grade). Table 1 lists the compounds in the samples used for the vulcanization imaging in thick rubber; all the reagents related to vulcanization except for filler materials used for high-sensitivity detection of mesh structure growth. The unit of "phr" in Table 1 is the preferred unit used in the rubber industry, indicating "parts per hundred rubber", which is the relative weight when the rubber polymer is 100. Samples for silica dispersion visualization had a different formulation, as shown in Table 2. The main difference from Table 1 was silica filler material, and a silane coupling reagent, which improves filler dispersion.

Table 1. Compounding formulation of the thick sample for vulcanization evaluation.

Ingredient	Material	Quantity [phr]
Polymer	SBR	100
Accelerator activators	Stearic Acid	1
	Zinc Oxide	5
Vulcanization accelerators	TMTD	1
	MBTS	1
	CBS	1
Vulcanizing agent	Sulfur	1

Materials included styrene-butadiene rubber (SBR), tetramethlthiuram disulfide (TMTD), dibenzothiathiazole disulfide (MBTS), and n-cyclohexyl-2-benzothiazole sulfenamide (CBS).

Table 2. Compounding formulation of silica sample.

Ingredient	Material	Quantity [phr]
Polymer	IR or SBR	100
Filler	Silica	30–130
Coupling agent	TESPT	Weight ratio of 8% of Silica
Accelerator activators	Stearic Acid	2
	Zinc Oxide	2.5
Vulcanization accelerators	DPG	1.5
	MBTS	2.5
	TBBS	0.7
Vulcanizing agent	Sulfur	1.6
Others	Wax	2
	Antioxidant	1

Materials included isoprene rubber (IR), styrene-butadiene rubber (SBR), diphenylguanidine (DPG), triethoxylpropyl tetrasulfide (TESPT), dibenzothiathiazole disulfide (MBTS), and N-(tert-butyl)-2-benzothiazole sulfenamide (TBBS).

The samples in Table 1; Table 2 were vulcanized via different methods. The sample for visualizing thick rubber curing was made with a mold, illustrated in Figure 1, that cured samples 35 mm in diameter and 20 mm thick. The vulcanization conditions were a pressure of 10 MPa and a temperature of 150 °C. For THz-TDS analysis, the cured thick rubber sample was sliced into six 2 – 4-mm-thick pieces, as shown in Figure 2. For THz absorbance spectra, 25 points in the central part of each sliced piece were evaluated. In the case of silica dispersion, masterbatch mixing, final mixing samples, and cured samples were used and imaged with a scanning electron microscope (SEM, CARL ZEISS AG, Ultra55), in addition to the THz-TDS characterization. The cure conditions were 10 MPa pressure at 160 °C. Here, we describe the experimental results of visualization of vulcanization and filler dispersion. These two conditions affect the quality of the rubber products considerably, and they are the main subject of concern of rubber manufactures; however, it was difficult to obtain their information rapidly and destructively using conventional techniques.

Filler dispersion was divided into two categories, micro-dispersion and macro-dispersion. Micro-dispersions were sub-μm-sized aggregates that could be observed by an electron microscope. Here, the emphasis was on macro-dispersions of mm–μm-sized aggregates that could be evaluated visually or with an optical microscope. However, an established method has not been developed for objective numerical evaluations. As previously reported [10], the THz-TDS technique has the potential to observe the macro-dispersions of CB filler. In this study, the silica macro-dispersion was visualized.

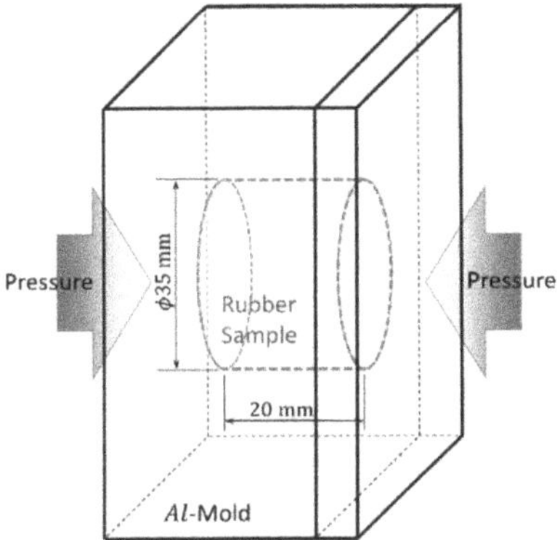

Figure 1. Mold for thick rubber production.

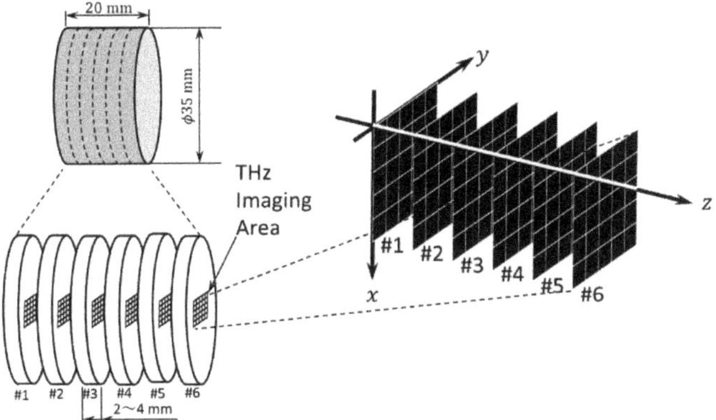

Figure 2. Sample preparation and measured areas for THz-TDS evaluation of thick rubber.

3. Data Processing

The THz-TDS data were evaluated as the absorbance A per unit thickness (cm) using

$$A(\nu) = \frac{\log \frac{I_0(\nu)}{I(\nu)}}{T} \left[\mathrm{cm}^{-1}\right] \tag{1}$$

where I_0 and I are the reference and transmitted THz light intensities through the sample, respectively, and T is the sample thickness. After the THz absorbance spectrum was obtained at position (x, y) on the sample, the representative value $A_{rep}(x,\ y)$ of the THz absorbance for the imaging process was calculated by

$$A_{rep}(x,\ y) = \int_{\nu_1}^{\nu_2} \frac{A_{xy}(\nu)d\nu}{\nu_2 - \nu_1} \tag{2}$$

where ν_1 and ν_2 were the low- and high-end frequencies of the THz absorbance spectrum, respectively. The statistical values, such as mean, standard deviation (SD), and coefficient of variation (CV), were derived by $A_{rep}(x,y)$ s for one sample. CV defined by SD/mean was useful for measuring the degree of uniformity of the $A_{rep}(x,y)$ values over the sample. The penetration depth of the rubber samples used in this study was estimated to be about 1 μm. The transmission loss due to Fresnel reflections from rubber-air interfaces was estimated to 4~5% per interface; however, all the results processed in this study were calculated without considering this loss. This is because it was difficult to obtain an accurate loss for each rubber sample, and it was expected that the magnitude of the loss was not so critical when the THz transmission was evaluated as absorbance having a logarithm scale.

To estimate the dispersion of the silica filler, a new parameter evaluating fluctuations in the THz absorbance spectral profiles was introduced. Usually, the coefficient of correlation is used for this purpose; however, differences of the coefficient were small when applied to fluctuations of the absorbance spectra. Therefore, spectral profile dissimilarity (*SPD*) was used instead. Specifically, nine or 25 points in each sample were measured to evaluate the filler dispersion, and the *SPD* numerically calculated differences in the spectra by

$$SPD = \frac{\sum^{n}_{\substack{j=1 \\ (j \neq k)}} \left[\frac{\sum_{i=1}^{m} \left| \frac{A_k(\nu_i) - A_j(\nu_i)}{A_k(\nu_i)} \right|}{m} / \left| r_{kj} \right| \right]}{n-1} \tag{3}$$

where $A_k(\nu)$ and $A_j(\nu)$ are the absorbances of the standard and compared spectra, respectively, as shown in Figure 3a. The r_{kj} is Pearson's coefficient of correlation, which is usually 0.9 for THz absorbance spectra. Once the *SPD* value for one particular standard spectrum was obtained, the next standard spectrum was selected to calculate the next *SPD*. If n points were sampled in one sample, the *SPD* s were averaged like,

$$SPD_{sample} = \sum_{k=1}^{n} \frac{SPD_k}{n} \tag{4}$$

Figure 3b shows an example of $n = 4$. Here, the *SPD* values were indicated as 100 times *SPD*, 100*SPD*. For complete overlaps between all spectra, 100*SPD* takes the null value.

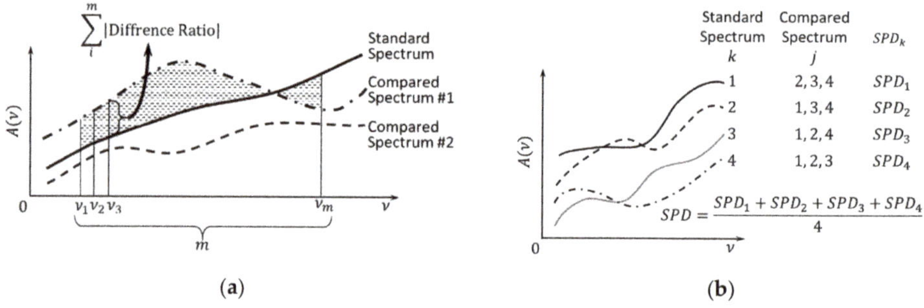

Figure 3. Definition of spectral profile dissimilarity (SPD). (**a**) SPD calculation, and (**b**) averaged SPD in one sample.

To confirm the filler dispersion evaluation via THz absorbance, the silica dispersions were estimated with SEM. In the rubber industries, the macro-dispersion of filler in SEM images is usually evaluated with number or weight averaging of aggregates, which is correctly cross-section weight aggregate cross-section. This cross-section is calculated by using processed SEM images where the aggregate cross-section thresholds were 1 μm^2 and 10 μm^2, for example. The filler dispersion was evaluated using a statistical parameter for an accurate estimation. The discrepancy D in the $k-$ dimensional unit cube is defined as [11,12]

$$D_N^{(k)} = \sup_{t \in [0,1]^k} \left| \frac{\#([0,t);N)}{N} - t_1 \times \cdots \times t_k \right|, \ \boldsymbol{t} = (t_1, \cdots, t_k), \quad (5)$$

where # is the counting function, N is the number of random nodes, and $\boldsymbol{t}$ is the point set. This value varies from 0 to 1, where smaller values indicate better dispersion. Because calculating this discrepancy is cumbersome even on a computer [12], a "simplified discrepancy" was used based on the concept of the discrepancy in this study. The idea of the new parameter is shown in Figure 4. The evaluated sample was equally divided into 16 areas to count the particle (filler) after image binarization. The ratio of filler number to the total filler number N in the sample was calculated, it was then subtracted by the normalized area section in each divided area, and its absolute value was calculated. Finally, the "simplified discrepancy" was obtained as the mean value over the 16 divided areas. The equation is as follows:

$$D_{simple} = \frac{\sum_i^n \left| \frac{\#(area_i)}{N} - S_{norm} \right|}{n}, \quad (6)$$

where n and S_{norm} are the divided area number in one SEM image and the normalized area cross-section, respectively. Here, the arithmetic mean was used instead of the operator "sup" in Equation (5), because the considered regions in one sample were only 16. This "simplified discrepancy" had values within a range of 0 to $(n-1)/n^2$, where 0 indicates complete uniformity.

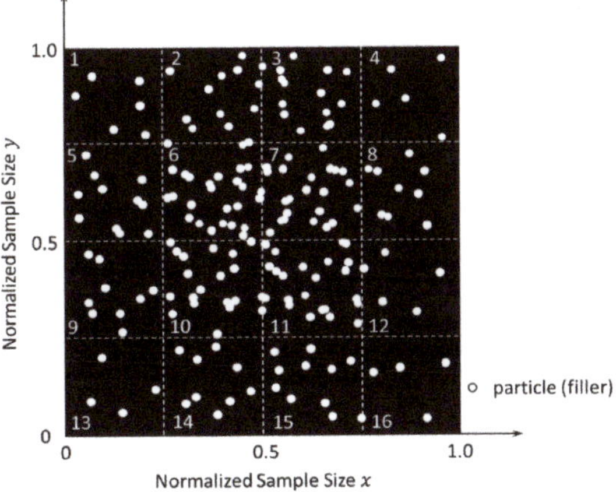

Figure 4. The concept of the simplified discrepancy calculation. This figure shows an example of $n = 16$.

4. Results and Discussion

4.1. Visualization of Vulcanization in Thick Samples

A simple analysis of vulcanization conditions in thick samples has been reported previously [10]. Prior to the analysis of vulcanization depth by the THz light, the dependence of THz absorbance on the cure time was investigated. This procedure yielded two cure times, *T*90 and *T*100, for visualizing vulcanization determined by a curemeter (JSR Trading Co., Ltd. (JSR Corporation), CURELASTOMER

W). $T90$ is generally called the "optimum vulcanization time," while $T100$ is the vulcanization finishing time. This simple analysis indicated that the central part of the sample had a low cure speed, while the THz absorbance in the center of the $T100$ sample was higher than that for $T90$. The analysis only considered heat sources for vulcanization located in the pressed direction, and the analysis target was six sliced pieces; however, the actual thick sample was surrounded by the aluminum mold shown in Figure 1. This means that the sample was heated from all directions, and the sliced pieces each had a distribution of vulcanization. To solve this problem, the analyzed areas were set as shown in Figure 5. The previously analyzed sliced areas were depicted as #1–#6 in Figure 5a. The averaged THz absorbance values in each sliced piece had been evaluated previously [10]. Here, a total of 150 measured points was classified into the three sections "around the center", the "central part", and the surroundings of each ("edge"), as shown in Figure 5b,c. Figure 6 compares the THz absorbance and the CV values of the THz absorbance in each section. The comparison results for "around the center" and the "edge" are shown in Figure 6a,b, while Figure 6c,d indicates the results for the central and the edge parts. For the $T90$ sample, the THz absorbance differences between the edge and "around the center" in Figure 6c are more significant than those in Figure 6a. The same tendency was observed in the CV comparisons in Figure 6b,d. Previous results showed that the THz absorbance was proportional to the cure progress [10]. Therefore, considering the distance from the heat source, the results here suggest that the vulcanization reaction in the edge part of the $T90$ sample was active and inhomogeneous, while in the "around the center"/central parts, the reaction progress was slower [10]. For the $T100$ sample, there was an almost equal THz absorbance for the edge and the "around the center"/central parts. This was understood by assuming that the cure reaction was nearly finished, and as a result, the CV values of the edge became small. Furthermore, the edge part in T100 samples indicated considerably larger CV values than that of the "around the center"/central parts. This result probably reflected that over-cure [10], which causes mesh structure destruction, occurred in T100 samples. Thus, it was confirmed that these detailed analyses revealed more information about vulcanization inside the thick rubber product.

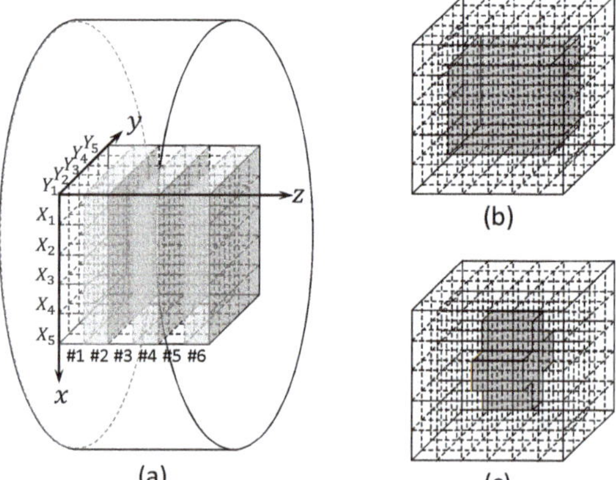

Figure 5. Thick rubber sample for detailed analysis in this study. (**a**) Coordinate axes and sliced areas (#1~#6) in the sample, (**b**) the "around the center" (36 points) and its surroundings ("edge," 114 points), and (**c**) the center (10 points) and its surroundings ("edge," 140 points).

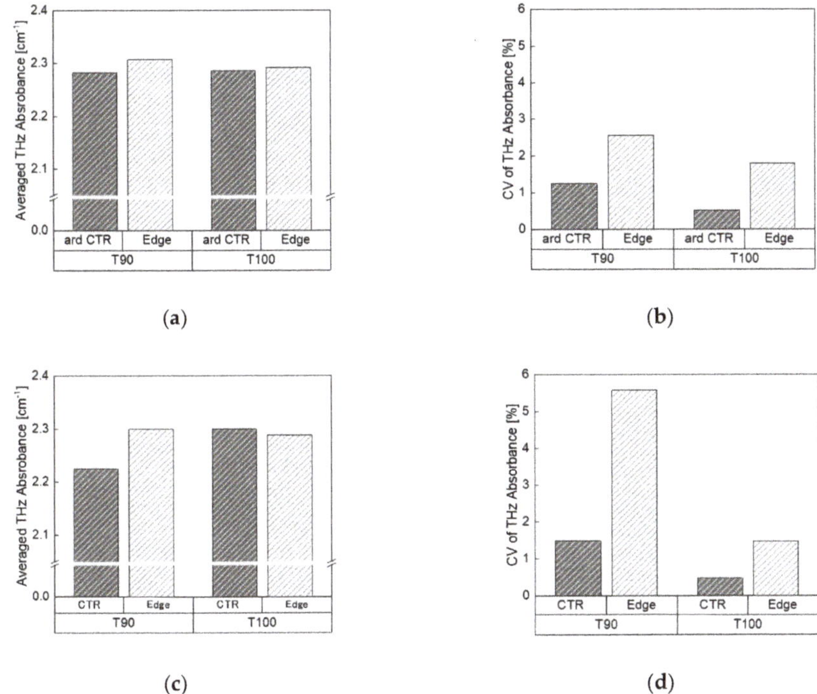

Figure 6. Comparison of the THz absorbance and the CV values of the THz absorbance in each section of *T*90 and *T*100 samples. (**a**) and (**c**) averaged THz absorbance in "around the center" and central part, and (**b**) and (**d**) CV of THz absorbance in "around the center" and central part, respectively.

4.2. Silica Dispersion in Rubber Samples

Previously the dispersion visualization of CB, the most general filler in rubber, was reported [10]. Here, silica was visualized because it is a significant filler in ecological automobile tires. Because several unmeasured additives were used in the samples, THz absorbance spectra were evaluated for each of the new additives, as shown in Figure 7. The identification of observed peaks was not performed in this study. By considering the amount of each reagent used in Table 2, it was found that the diphenylguanidine (DPG), antioxidant, and wax were negligible in the THz absorbance of the rubber sample as much as other vulcanization accelerators and vulcanizing agent. Silane is an essential chemical material when silica is used as a filler in rubber products because silica cannot be uniformly distributed due to silanol groups on the silica surface. The silane bis(3-Triethoxylpropyl) tetrasulfide (TESPT in Table 2) reacted with silica at temperatures less than 140 °C, shown in Figure 8 [13,14]. The THz absorbance change due to this reaction was monitored in a preliminary experiment that used powdered samples. Because the bulk density of the powdered silica was approximately one-half that of the tablet silica, in addition to a scattering of THz light by powdered silica, the measured absorbance was smaller than that in Figure 7. When liquid silane was mixed with the silica, the silica tended to clump, and the absorbance spectrum was obtained by averaging three points. In contrast, the absorbance of the pure silica powder sample was stable. After heating at 130 °C for 2 h, the spectra of the mixed powder were as stable as those of the pure silica; however, the THz data were measured at the different three points in the sample and averaged. From Figure 9, it was found that the absorbance increased with silane addition and, after heating (silica-silane reaction), the absorbance decreased. Although the cause of this decrease was unclear (it may be an intermediate product, shown in Figure 8), the results indicated the detection of the silica inside rubber samples.

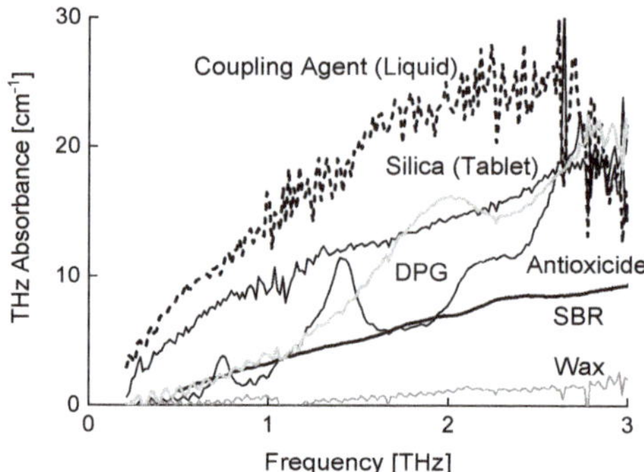

Figure 7. THz absorbance spectra for newly measured agents.

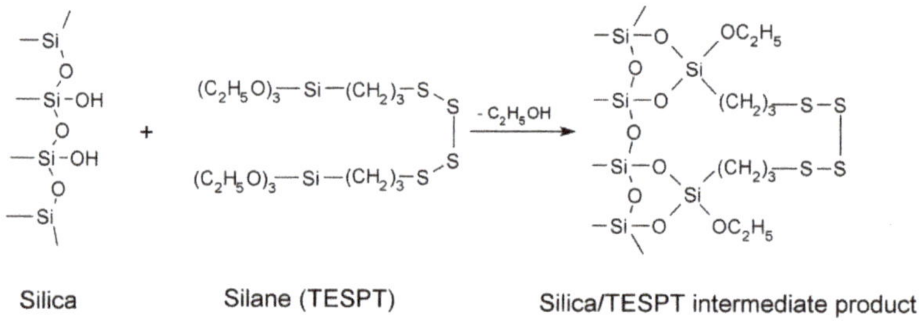

Figure 8. Chemical reaction between silica and silane.

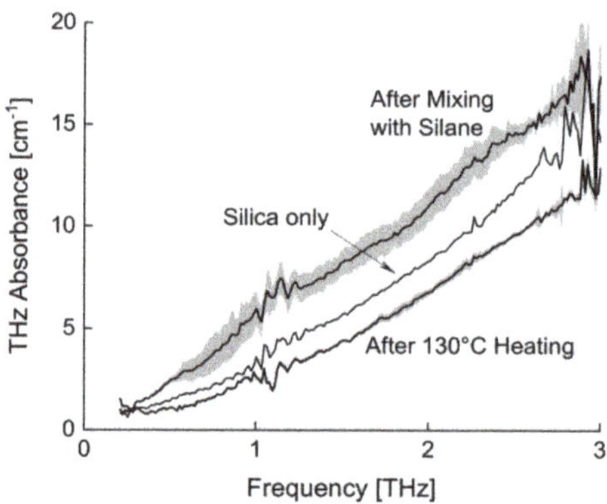

Figure 9. Change of the THz absorbance spectrum by silica-silane chemical reaction. Gray-colored zones mean standard errors.

A calibration curve of the silica THz absorbance was used to visualize the dispersion of silica filler inside the rubber samples. Samples made from two different polymers were prepared because SBR polymer, which is the main target and widely used in automobile tires, has a lower viscosity and affects filler dispersion. Isoprene rubber (IR) was selected for comparison because it has a simple structure, a higher Mooney viscosity (stiffer), and an almost equal THz absorbance relative to SBR. When the calibration curves were calculated for silica filler, the same procedure was chosen as that for CB filler dispersion [10]. The THz-TDS system evaluated 25 points in each sample and processed the data to produce absorbance images. The results are shown in Figure 10. Both the IR and SBR samples had excellent correlations between the THz absorbances and the amount of silica. The squared Pearson's coefficient of correlation, r^2, was 0.994 and 0.990 for IR and SBR, respectively. Comparing the previous results using CB filler, no saturation was observable at higher silica amounts. This suggests that internal scattering or multiple scattering inside the rubber samples was not dominant even in samples having higher filler concentration. This assumption was supported by the fact that remarkable temporal THz waveform degradation [15] was not observed in samples containing silica filler (the waveform is not shown). The parameters of *CV* and *SPD* were plotted together in Figure 10. The *CV* was obtained from the integrated absorbance values, which eliminated the THz absorbance spectral fluctuations. In contrast, the *SPD* reflected the difference in spectral profiles. Increased filler concentration induced more uniform dispersion; thus, the *CV* and *SPD* trends in Figure 10 were acceptable. Furthermore, the variations were comparable. Both samples indicated minimum values at 70 phr silica; however, the cause was not understood. The 70-phr silica possibly matched the quantities of other additives and polymer weights. Figure 10 verified that THz absorbance was suitable for evaluating the silica dispersion, and that the parameter *SPD* was a potential indicator of spectral uniformity.

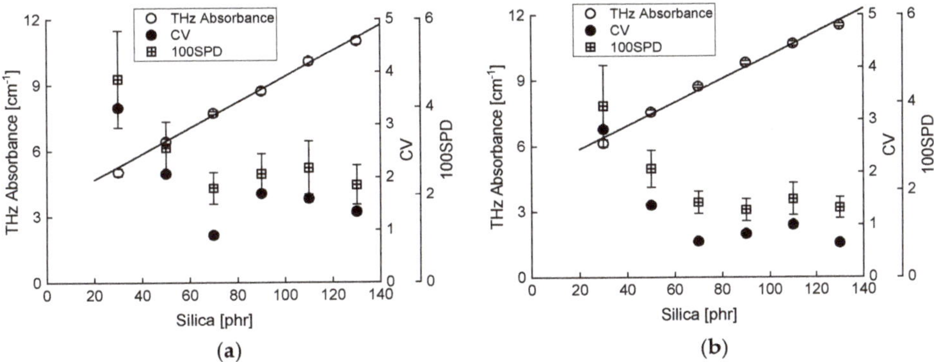

Figure 10. Calibration lines for the THz absorbance (solid line) against silica concentration with *CV* of the THz absorbance and 100*SPD* of the THz absorbance spectra. (**a**) IR- and (**b**) SBR-based sample.

The visualization of silica dispersion was performed using the same imaging procedure as that used for CB filler [10]. The grayscale images for the silica dispersion were inferior to the CB dispersion images in recognition of the dispersion. Meanwhile, the fluctuation of the absorbance spectrum was suitable to evaluate silica dispersion in the samples. This was because there was much less silica or silica-silane compound absorption relative to that of CB, although the linearity of the silica absorbance was confirmed in Figure 10. The 25 spectra were overlapped in one plot in Figure 11 to identify fluctuations more easily. The THz absorbance image is shown in the inset of each spectrum for reference. The spectra were for (a) silica masterbatch mixing (unvulcanized), (b) final mixing (unvulcanized), and (c) vulcanized rubber. These production processes of rubber products, such as masterbatch mixing and final mixing, were explained previously [10]. The first process was the masterbatch mixing, and the last condition of the product was a vulcanized rubber, which follows the final mixing. As shown in Figure 11, the fluctuations settled as the procedure progressed. The *SPD* in Equations (3) and (4) was used to evaluate the spectral fluctuations. Table 3 summarizes the averaged absorbance, *CV*, and 100*SPD* for these three conditions over the range 1.2–2.6 THz. The mean value of the absorbance was almost constant, while the fluctuations gradually decreased with the process. To confirm the validity of the fluctuation evaluation with the *SPD* parameter, SEM images were acquired for samples of masterbatch mixing and cured rubber. The SEM images and the analysis of dispersion by the simplified discrepancy are shown in Figure 12. The brightness of the images was slightly adjusted by software to emphasize the surface conditions for the (a) masterbatch mixing and (b) cured rubber. In Figure 12b, the vulcanized rubber had a clean surface. The evaluation by the simplified discrepancy is shown in Figure 12c, which has two bars for the masterbatch mixing sample (from two SEM images) and one bar for the cured rubber. The data confirmed that the value of the simplified discrepancy for the cured sample was smaller with fewer errors than that of the masterbatch sample, and that the dispersion evaluation of silica filler by THz absorbance was valid. This result proved that the novel parameter of the simplified discrepancy can function for the SEM image analysis.

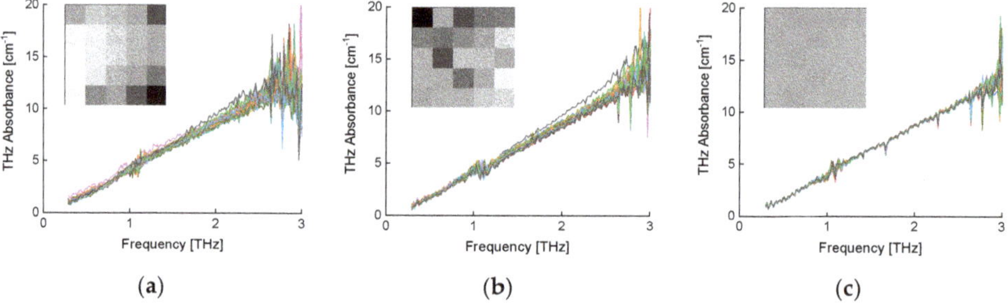

Figure 11. THz absorbance spectra at 25 measured point in various sample condition. (**a**) Silica masterbatch mixing (unvulcanized), (**b**) final mixing (unvulcanized), and (**c**) vulcanized rubber. Inset shows each THz absorbance image.

Table 3. The averaged THz absorbance, CV, and 100SPD for masterbatch mixing, final mixing, and vulcanized rubber.

Sample	Masterbatch Mixing	Final Mixing	Cured Rubber
Mean	8.20	8.32	8.22
CV [%]	3.04	2.84	0.242
100*SPD*	6.32	5.68	2.38

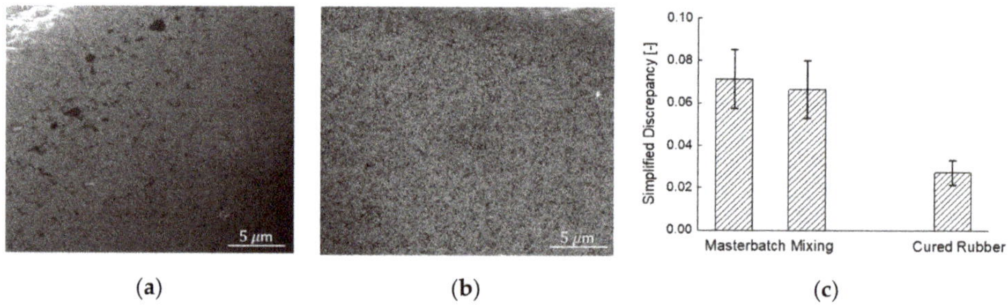

Figure 12. SEM image evaluation of SBR-based sample. (**a**) Silica masterbatch mixing, (**b**) vulcanized rubber, and (**c**) silica dispersion in SEM images derived by simplified discrepancy.

5. Conclusions

The vulcanization condition inside thick rubber samples was visualized by THz absorbance images in detail. They revealed that the surrounding aluminum mold acted as a heat source, and that the central and the "around the center" regions had a lower rate of vulcanization relative to the edges. The THz absorbance characterized the dispersion of the silica filler in the rubber samples, which is difficult to evaluate nondestructively and rapidly with other methods. The profile fluctuation of the absorbance spectrum was an effective indicator of the homogeneity of the silica filler. The validity of this method was confirmed with SEM imaging, where a new parameter was used to evaluate the filler uniformity. Overall, the THz technique could be used for detecting or assessing the distribution of silica, which will have more importance in future rubber products.

Author Contributions: Conceptualization, M.H., T.G and Y.H.; methodology, T.G. and Y.H.; software, Y.H.; validation, M.H., T.G. and Y.H.; formal analysis, Y.Y., T.M. (Toshiaki Morichika), R.S., T.M. (Takakazu Minato) and Y.H.; investigation, Y.Y., R.S., T.M. (Toshiaki Morichika), T.M. (Takakazu Minato) and Y.H.; resources, M.H., T.G and Y.H.; data curation, M.H., T.G., Y.Y. and Y.H.; writing—original draft preparation, Y.H.; writing—review and editing, Y.Y., T.G., R.S., T.M. (Toshiaki Morichika), T.M. (Takakazu Minato), M.H. and Y.H.; visualization, Y.H.; supervision, M.H. and Y.H.; project administration, M.H. and Y.H.; funding acquisition, M.H. All authors have read and agreed to the published version of the manuscript.

Funding: This research received no external funding.

Acknowledgments: The authors thank associate professor Katsunori Watanabe and technical staff member Takuta Kamino, the national institute of technology (KOSEN), Kurume College, for helping in making the rubber compounds. We thank Alan Burns, from the Edanz Group (www.edanzediting.com/ac) for editing a draft of this manuscript.

Conflicts of Interest: The authors declare no conflict of interest.

References

1. Tonouchi, M. Cutting-edge terahertz technology. *Nat. Photonics* **2007**, *1*, 97–105. [CrossRef]
2. Schmuttenmaer, C.A. Exploring Dynamics in the Far-Infrared with Terahertz Spectroscopy. *Chem. Rev.* **2004**, *104*, 1759–1780. [CrossRef] [PubMed]
3. Heilweil, E.J.; Plusquellic, D.F. Terahertz Spectroscopy of Biomolecules. In *Terahertz Spectroscopy Principles and Applications*; Dexheimer, S.L., Ed.; CRC Press: Boca Raton, FL, USA, 2007; pp. 269–298.
4. Hoshina, H.; Hayashi, A.; Miyoshi, N.; Miyamaru, F.; Otani, C. Terahertz pulsed imaging of frozen biological tissues. *Appl. Phys. Lett.* **2009**, *94*, 123901. [CrossRef]
5. Rungsawang, R.; Geethamma, V.G.; Parrot, E.P.J.; Ritchie, D.A.; Terentjev, E.M. Terahertz spectroscopy of carbon nanotubes embedded in a deformable rubber. *J. Appl. Phys.* **2008**, *103*, 123503. [CrossRef]
6. Piesiewicz, R.; Jansen, C.; Wietzke, S.; Mittleman, D.; Koch, M.; Kürner, T. Properties of Building and Plastic Materials in the THz Range. *Int. J. Infrared Millim. Waves* **2007**, *28*, 363–371. [CrossRef]
7. Yasui, T.; Yasuda, T.; Sawanaka, K.; Araki, T. Terahertz paintmeter for noncontact monitoring of thickness and drying progress in paint film. *Appl. Opt.* **2005**, *44*, 6849–6856. [CrossRef] [PubMed]
8. Fitzgerald, A.J.; Cole, B.E.; Taday, P.F. Nondestructive analysis of tablet coating thicknesses using terahertz pulsed imaging. *J. Pharm. Sci.* **2005**, *94*, 177–183. [CrossRef] [PubMed]
9. Hirakawa, Y.; Ohno, Y.; Gondoh, T.; Mori, T.; Takeya, K.; Tonouchi, M.; Ohtake, H.; Hirosumi, T. Nondestructive Evaluation of Rubber Compounds by Terahertz Time-Domain Spectroscopy. *J. Infrared Millim. Terahertz Waves* **2011**, *32*, 1457–1463. [CrossRef]
10. Hirakawa, Y.; Yasumoto, Y.; Gondo, T. Evaluation of Rubber Products by Terahertz Time-domain Spectroscopy Carbon Black Dispersion and Vulcanization State. *J. Infrared Millim. Terahertz Waves* **2020**. [CrossRef]
11. Niederreiter, H. Quasi-Monte Carlo Methods for Numerical Integration. In *Random Number Generation and Quasi-Monte Carlo Methods*; Society for Industrial and Applied Mathematics: Philadelphia, PA, USA, 1992; pp. 13–22.
12. Gandar, B.; Loosli, G.; Deffuant, G. Sample Dispersion Is Better than Sample Discrepancy for Classification, hal-00679061. 2010. Available online: https://hal.archives-ouvertes.fr/hal-00679061 (accessed on 1 February 2020).
13. Rodgers, B.; Wanddell, W. The Science of Rubber Compounding. In *The Scientific and Technology of Rubber*, 4th ed.; Mark, J.E., Erman, B., Roland, C.M., Eds.; Academic Press: Amsterdam, The Netherlands, 2013; pp. 417–471.
14. Ichino, T. Silane Coupling Agent for Rubber Usage. *J. Soc. Rubber Sci. Technol. Jpn.* **2009**, *82*, 67–72. [CrossRef]
15. Hirakawa, Y.; Yamauchi, T.; Kamino, T.; Gondoh, T.; Hirano, S.; Noguchi, T. Dependence of THz Signals on Carbon Black Compounding Amount in Vulcanized Rubber. In *Technical Digests of the Conference on Lasers and Electro-Optics (CLEO)*; Optical Society of America: San Jose, CA, USA, 2017.

Article

Nondestructive Testing of Hollowing Deterioration of the Yungang Grottoes Based on THz-TDS

Ju Feng [1], Tianhua Meng [2,*], Yuhe Lu [2], Jianguang Ren [3], Guozhong Zhao [4], Hongmei Liu [2], Jin Yang [1] and Rong Huang [2]

1 School of Geophysics and Information Technology, China University of Geosciences, Beijing 100083, China
2 School of Physics and Electronics Science, Shanxi Datong University, Datong 037009, China
3 The Research Institute of Yungang Grottoes, Datong 037007, China
4 Department of Physics, Capital Normal University, Beijing 100048, China
* Correspondence: mengtianhua@sxdtdx.edu.cn; Tel.: +86-1399-432-3503

Received: 24 February 2020; Accepted: 7 April 2020; Published: 9 April 2020

Abstract: Terahertz (THz) spectroscopy is an important method in noninvasive detection and diagnosis for historic relics. A new nondestructive testing (NDT) method based on terahertz time-domain spectroscopy (THz-TDS) technology was developed to measure the hollowing deterioration of the Yungang Grottoes in this paper. Hollowing deterioration samples were strictly prepared, and a series of experiments were conducted to ensure the representativeness of the experimental results. A hollowing thickness model was established by the relationship between the thickness of the hollowing deterioration sample and the time difference of the front flaked stone surface and the stone wall surface of the hollowing deterioration samples. The results show that the R-squared value of the model equation reached 0.99795, which implies that this model is reliable. Therefore, the actual hollowing thickness of the Yungang Grottoes can be obtained by substituting the time difference in the proposed thickness hollowing model, where the time difference is obtained from measured THz spectra. The detection method of stone relic hollowing deterioration is easy to apply, which can not only realize qualitative NDT but also quantitative hollowing deterioration thickness determination. This method has crucial practical significance for the repair and strengthening of stone relics similar to the Yungang Grottoes.

Keywords: terahertz spectroscopy; open stone relics; hollowing; weathered; preservation of cultural heritage

1. Introduction

Stone relics are monuments consisting of natural stones, including buildings, grottoes, stone tablets, etc., which have very high historical and artistic values, as well as represent a specific kind of geological engineering requiring long-term preservation. However, with the intensification of environmental pollution and under the influence of natural and human factors, stone relics (especially open stone relics) are being damaged at an alarming rate [1–3]. This damage has endangered the safety of cultural relic preservation and has affected the historical and cultural values of stone relics.

As one of the commonly occurring types of damage, hollowing deterioration destroys the structure and shortens the life of stone relics. Hollowing possesses concealed characteristics that cannot be observed timely and clearly from the stone surface. With the Yungang Grottoes as an example, as shown in Figure 1, the surface of stone relics is corroded, and the internal layered or sheet-like clay mineral and gypsum components produce a high expansion pressure under the action of natural precipitation, such as rain. When a rock formation has low rock strength, a hollowing structure will be produced, which comprises three parts, namely, the front rock wall (flaked stone), the air layer, and the substrate layer (stone wall). When the air layer in the hollowing structure reaches a certain thickness, the front rock wall may naturally detach, forming new cracks as a result of large temperature fluctuations and an uneven internal force distribution, which significantly endangers stone relics.

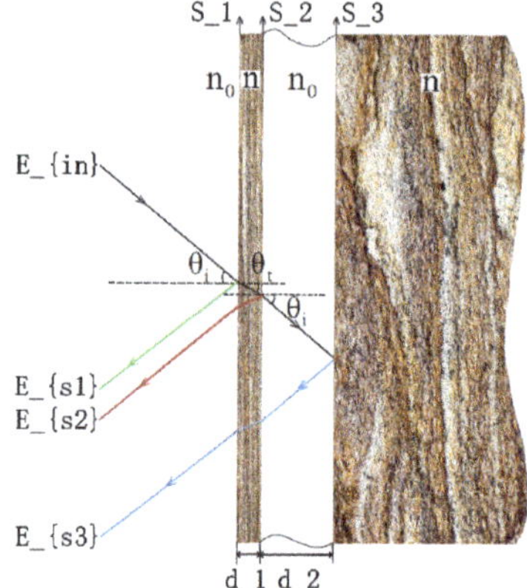

Figure 1. Schematic illustration of hollowing deterioration of stone relics and the terahertz (THz) wave reflection single-point thickness extraction principle of the hollowing model. E_{in} is the incident field; E_{s1}, E_{s2}, and E_{s3} are the fields reflected from the front and back flaked stone surfaces and the stone wall surface, respectively; d_1 and d_2 are the flaked stone and hollowing sample thicknesses, respectively; θ_i and θ_t are the incidence and refraction angles, respectively; and n and n_0 are the refractive indexes of the flaked stone sample and air, respectively.

The nondestructive detection technique is an effective method for identifying hidden internal defects and dislocation impurities to estimate and extend the life of the tested object by the use of sound, light, magnetic field, electricity, etc., under the precondition of not damaging the detection target [4–14]. Compared with the conventional methods in the present nondestructive testing (NDT) field, such as magnetic particle testing, penetration testing, eddy current testing, radiographic testing, and ultrasonic testing, the typical wavelength of terahertz (THz) waves (300 μm) is larger than the size of small-scale structures; therefore, THz scattering in most objects occurs far less than for visible and near-infrared light, and the THz wave photon energy is commonly lower than the chemical bond energy, which means that THz waves can better penetrate most nonpolar materials. Furthermore, a THz wave is particularly suitable for NDT, and it allows the development of non-destructive, non-contact, non-ionizing methods that could advantageously replace other evaluation methods based on X-rays, ultrasound, and thermography [15–21].

Terahertz time-domain spectroscopy (THz-TDS) is a new THz spectrum measurement technique based on ultra-short pulse technology developed in recent years, which has been widely applied in the fields of physics, chemistry, biology, etc. [22–27]. The physical and chemical parameters of tested objects, such as the complex dielectric constant, dispersion coefficient, transmission, and absorption, are usually obtained by THz-TDS, and the material composition and structure of the tested objects can then be studied. However, no relevant report is found on characterizing stone relic hollowing deterioration with THz-TDS. In this paper, the hollowing deterioration of Yungang Grotto samples was studied using THz nondestructive testing (THz-NDT) technology based on THz-TDS.

2. Hollowing THz-NDT Theory

THz-TDS is a coherent detection technology that can simultaneously obtain information about the THz pulse amplitude and phase [28–30]. As such, it relies on THz wave reflection to detect changes in the THz time-domain pulse wave in the sample before and after irradiation, called the reference and sample waveforms, respectively. The THz time-domain spectra of reference and hollowing deterioration samples are shown in Figure 2. The reference wave is the THz wave in the sample without hollowing deterioration, with a corresponding d_2 value of 0. When the corresponding d_2 values of the hollowing deterioration sample waves are 1, 2, 3, and 4 mm, the upper right figure shows that the THz time-domain spectra of the reference and hollowing deterioration samples range from 368 to 404 picoseconds. Figure 2 reveals that the different peak positions of the echo waves correspond to various prolonged times of hollowing samples with different thicknesses (d_2). With increasing hollowing deterioration sample thickness, the delay time of the echo wave increases. The peak positions are reflected from the front and back flaked stone surfaces (S_1 and S_2, as shown in Figure 1) corresponding to 314.88 and 374.17 ps, respectively. The peak positions reflected from the stone wall surface (S_3 and d_2, as shown in Figure 1) occur at 378.49, 384.45, 389.79, and 396.08 ps, which correspond to d_2 values of the hollowing deterioration model samples of 1, 2, 3, and 4 mm, respectively.

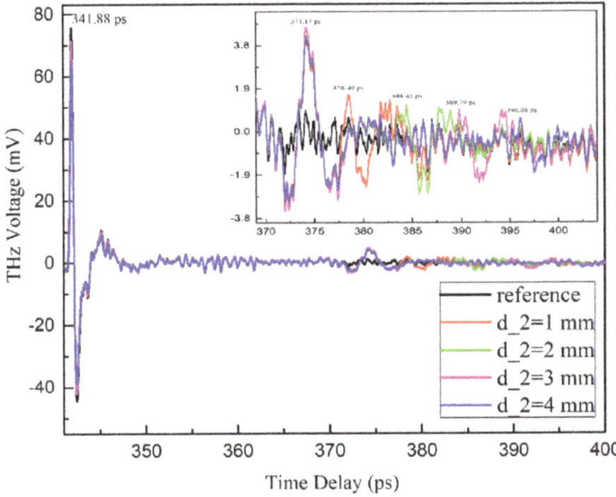

Figure 2. THz time-domain spectra of the reference and hollowing deterioration samples.

2.1. Single-Point Thickness Extraction Hollowing Model

THz waves are reflected at the interfaces between media with different dielectric constants during propagation, as shown in Figure 1. The THz wave reflection single-point thickness extraction principle of the hollowing model is based on the assumption that flaked stone is homogeneously and isotropically distributed at a scale that is relatively larger compared with the focal spot size of the THz wave.

When the THz wave has an incident angle θ_i, E_{s1} is the THz wave reflected by the front flaked stone surface (S_1), and T_1 is the peak position of the first wave, E_{s1}. Similarly, T_2 and T_3 are the peak positions of the first wave reflected by the back flaked stone surface (E_{s2}) and the stone wall surface (E_{s3}), respectively (as shown in Figure 1). According to the THz wave propagation theory, the thickness was defined in the reflection single-point extraction model as follows [13]:

$$\mathrm{d\_2} = \frac{c\sqrt{n_0^2 - n^2\sin^2\theta_t}}{2n_0^2}(\mathrm{T\_3} - \mathrm{T\_1}) - \frac{\sqrt{n_0^2 - n^2\sin^2\theta_t}}{\sqrt{n^2 - n_0^2\sin^2\theta_i}}\left(\frac{n^2}{n_0^2}\cdot \mathrm{d\_1}\right), \tag{1}$$

When the incident direction of the THz wave is perpendicular to the hollowing deterioration samples, the thickness of the single-point extraction model can be simplified as:

$$\mathrm{d\_2} = \frac{c}{2n_0}(\mathrm{T\_3} - \mathrm{T\_1}) - \frac{n}{n_0}\cdot \mathrm{d\_1}, \tag{2}$$

where d_1 and d_2 are the thicknesses of the flaked stone and the hollowing deterioration sample, respectively, T_1 and T_3 are the peak positions of the first wave reflected by E_{s1} and E_{s3}3, respectively, n and n_0 are the refractive indexes of the flaked stone sample and air, respectively, and c is the light propagation velocity in air.

2.2. Simplified Hollowing THz-NDT Model

The hollowing thickness can be obtained by measuring the THz echo time difference when the refractive index of the flaked stone is known. However, in a general engineering test, the optical parameters of samples are very difficult to extract. However, the THz echo time difference can represent the hollowing thickness, so we simplified the proposed model with reflective thickness correlation coefficients k and b as follows:

$$d\_2 = a \times (T\_3 - T\_1) + b = k \times \Delta T + b, \tag{3}$$

Currently, irregularly shaped targets cause challenges in hollowing deterioration detection in engineering applications. Therefore, the simplified hollowing thickness model is pre-established by the correlation coefficient method to solve this problem. The actual hollowing thickness of the Yungang Grottoes can be determined by substituting the time difference in the hollowing thickness model, and this difference time can be obtained from measured THz spectra.

3. Experimental Method

3.1. Nondestructive Testing Using THz Wave Technique

In this paper, the hollowing deterioration samples are tested with a THz-TDS1008 test system, which is compact, self-contained, and highly integrated, and the optical antenna method is adopted to produce and detect THz pulses. The latter approach relies on a femtosecond laser, THz emission and detection components, and a composite time-delay system. The central wavelength of the laser is 800 nm, the pulse duration is 100 fs, the THz pulse width is 0.05~3.5 THz, and the signal-to-noise ratio (SNR) >65 dB. In addition, the hollowing deterioration model samples were tested at normal temperature (293 K) and 30% relative humidity. The schematic experimental setup is shown as Figure 3.

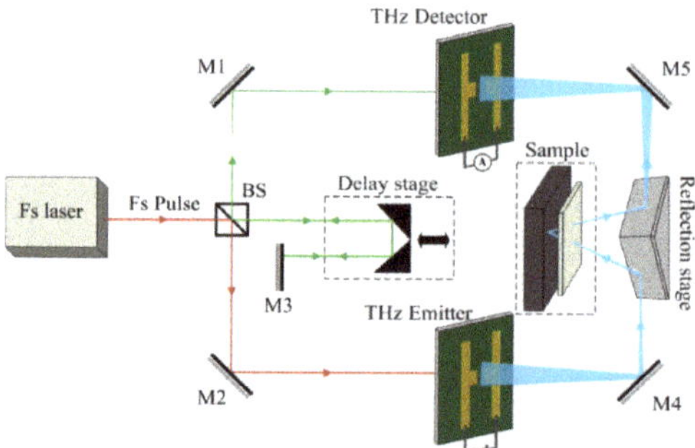

Figure 3. Schematic of the experimental setup. BS: splitter; M: mirror.

3.2. Hollowing Deterioration Sample Preparation and Testing

Hollowing deterioration is a common stone surface degradation phenomenon, in which surface sheet-like layers experience uplift deformation leading to cavities and cracks. Over time, these layers will detach due to their weight and the environmental changes, such as temperature, humidity, shock, and vibration. The THz-TDS can be effectively applied to the quantitative diagnosis of hollowing deterioration. In this work, we have avoided the inhomogeneous samples so that precise THz spectra can be obtained. We cut grotto samples into approximately 2-mm stone sheets and 6-mm stone blocks using an angle grinder and then polished them with sandpaper, which were used as the stone flake and stone wall of the hollowing deterioration model. Finally, we overlaid the two stone slices with different thicknesses to form a hollowing deterioration sample, as shown in Figure 4. The hollowing deterioration model samples were tested, and the thickness range of the hollowing deterioration samples was 0 to 4 mm at 0.1-mm intervals.

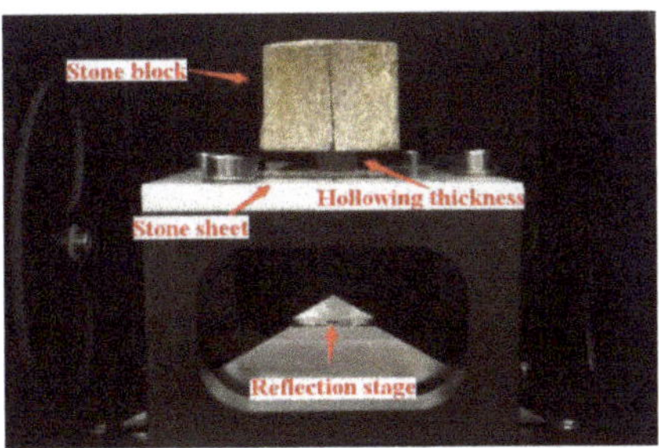

Figure 4. Schematic of the hollowing deterioration sample.

Since the thickness of the front surface of most hollowing deterioration in Yungang Grottoes is about 2 mm, we chose this typical thickness for our study. At the same time, we also tested the time-domain spectra of the flaked stone in the THz reflection and transmission system, as shown in Figure 5. Figure 6 is the refractive index spectrum. As can be seen from Figure 6, the refraction index of flaked stone was 2.11 in the THz band.

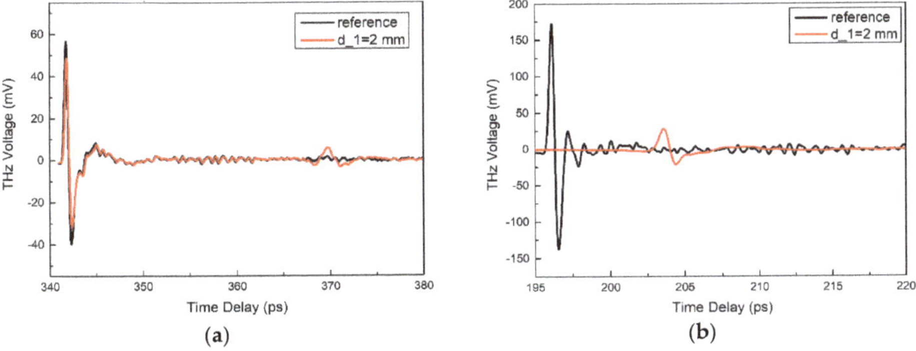

Figure 5. THz time-domain spectra of the reference and the flaked stone thickness of 2 mm. The time-domain spectra of (**a**) were obtained from the THz reflector system, with the transmission THz system for (**b**).

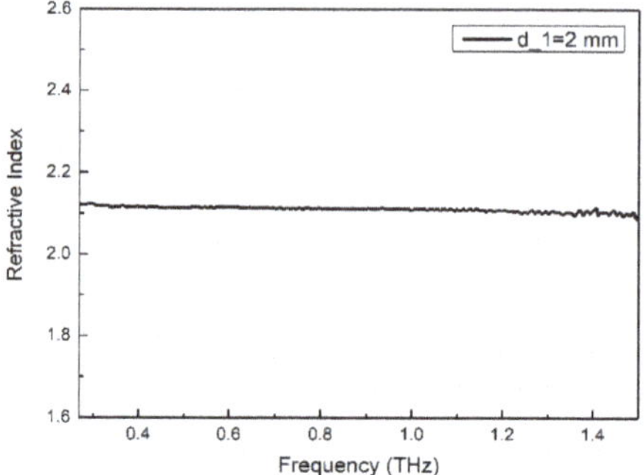

Figure 6. The THz refractive index spectra of the flaked stone thickness of 2 mm.

4. Results and Discussion

To accurately and reliably obtain the hollowing thickness, we measured THz spectra of the hollowing deterioration samples, where the thickness ranged from 0 to 4 mm at 0.1-mm intervals, and the measured data was used to develop the hollowing detection model. The THz time-domain spectra for all the hollowing deterioration samples are shown in Figure 7. Figure 7a–d are the THz time-domain spectra of the hollowing deterioration samples with different thicknesses. As THz wave is very sensitive to the small changes of the hollowing deterioration samples, the THz time-domain spectra of the different hollowing deterioration samples were different in the terahertz band and could be distinguished. In addition, it should be pointed out that the stone sheet of the hollowing deterioration samples was relatively thin but had a high refractive index. So the oscillating wave mainly resulted from the multiple reflections inside the sample, which is the Fabry–Perot interference effect. THz time-domain spectra revealed significant differences among the hollowing deterioration samples with different thicknesses, which caused the different propagating velocities in sample paths that gave rise to the different time delays. The decrease in terahertz pulse intensity was due to the reflectivity and absorption of the sample and the THz pulse became broad with the dispersion of the sample. Figure 7 indicates that the THz spectra of all samples attenuated quickly with the increasing of time delay in the whole spectral regions. The thickness of hollowing deterioration samples varied linearly with the peak position of the first wave reflected by the stone wall surface of the hollowing deterioration samples (T_3).

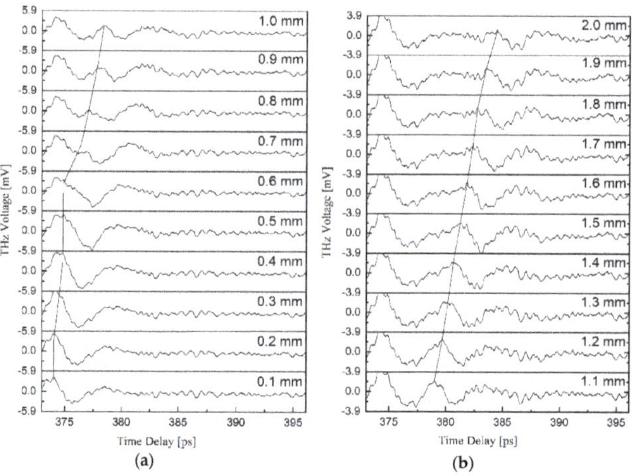

Figure 7. *Cont.*

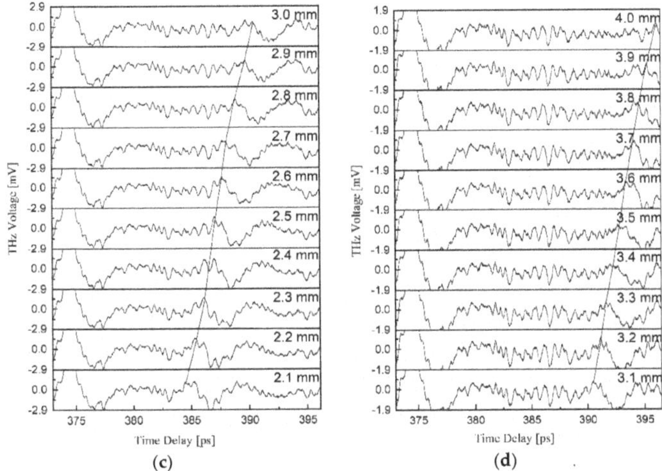

Figure 7. THz time-domain spectra of the hollowing deterioration samples. The thickness of the hollowing deterioration samples in (**a**) are 0.1 to 1.0 mm at 0.1-mm intervals. The thickness of the hollowing deterioration samples (**b**) are 1.1 to 2.0 mm at 0.1-mm intervals. The thickness of the hollowing deterioration samples (**c**) are 2.1 to 3.0 mm at 0.1-mm intervals. The thickness of the hollowing deterioration samples (**d**) are 3.1 to 4.0 mm at 0.1-mm intervals.

The peak positions of the THz spectra (T_3) for all of the different thicknesses were extracted with the hollowing model, and the reflection peak position of the S_1 surface (T_1, 314.88 ps) and the S_2 surface (T_2, 374.16 ps) was then subtracted to obtain the THz time delay difference (ΔT). As the thickness of the air layer in the hollowing deterioration increased, several close echoes appeared at the same time. Grubbs' test was selected to eliminate the noise value when T_3 was selected accurately in the measured THz spectra. By Grubbs' table look-up method, we were able to obtain the values of G (n_0) for use in excluding outliers (for which G is greater than G (n_0)). G is defined as follows:

$$G = \left|\frac{\bar{t} - t_n}{s}\right| = \left|\frac{\frac{1}{n}\sum_{i=1}^{n} t_i - t_n}{\sqrt{\frac{1}{n-1}\sum_{i=1}^{n}\left(t_i - \frac{1}{n}\sum_{i=1}^{n} t_i\right)^2}}\right| \tag{4}$$

where $\bar{t}$ is the average of all THz wave time (t) data for every sample, s is the standard deviation, and n_0 is the significance level, which was taken to be 5%.

Figure 8 shows the relationship between ΔT and d_2, and in addition, the experimental data were linearly fitted. The R-squared value was up to 0.99795, which verifies the feasibility and validity of this method. The linear fitting equation for ΔT and d2 can also be described as follows:

$$d\_2 = -9.88215 + 0.17121 \times \Delta T \tag{5}$$

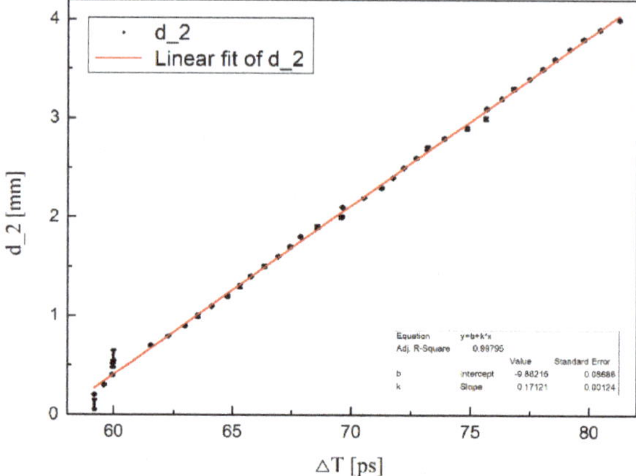

Figure 8. The relation curve between d_2 and ΔT and the linear fit of d_2.

By comparing the above linear fitting equation and Equation (3) of the hollowing thickness model, the correlation coefficients b and k of the latter could be determined as 9.88215 and 0.17121, respectively. In a general survey of Yungang Grottoes deterioration, the accurate hollowing thickness can be calculated by substituting actually measured THz spectra in the proposed model, which can provide effective references for repairing and reinforcing ancient relics.

5. Conclusions

In this work, we established a hollowing deterioration thickness detection model based on THz-TDS single-point experimental measurements. Our analysis suggests that THz technology can be applied to efficiently detect hollowing deterioration, as it reflects the thickness of hollowing deterioration. Even though the refractive index of hollowing deterioration samples is unknown, the model is universal. In the case of the hollowing deterioration samples of Yungang Grottoes, the THz spectra of 40 hollowing deterioration samples were determined, and the linear relationships between the thickness of hollowing deterioration samples and THz wave time-delay differences of the front flaked stone surface and the stone wall surface in samples were investigated. The resulting statistical model R-squared value reached 0.99795, which verified the feasibility and validity of this model. The method of analyzing the hollowing thickness of cultural relics using the THz-TDS method, which has high precision, is simple and nondestructive, and the method can be used for real-time hollowing deterioration detection of the Yungang Grottoes and be extended to other cultural relics. Moreover, the development of miniaturized, integrated, and higher-resolution THz instrumentation will enable this method to be applied in the field of cultural relic detection.

Author Contributions: Conceptualization, J.F., T.M., Y.L., G.Z., H.L., and J.Y.; Data curation, J.F., T.M., J.R., and R.H.; Investigation, J.F., T.M., and J.R.; Methodology, J.F., T.M., G.Z., and R.H.; Supervision, Y.L., G.Z., and J.Y.; Validation, J.F. and T.M.; Visualization, J.F., T.M., Y.L., H.L., J.Y., and R.H.; Writing—original draft, J.F. and T.M.; Writing—review and editing, J.F., T.M., G.Z., H.L., and J.Y. All authors have read and agreed to the published version of the manuscript.

Funding: This work was supported by the National Natural Science Fund of China under grants 11504212, the Science and Technology Innovation Group of Shanxi Province, China under grants 201805D131006, Important R&D projects of Shanxi Province under grants 201803D121083 and 201803D31014, Applied Basic Research projects of Shanxi Province under grants 201801D121072, and the Ph.D. research startup foundation of Shanxi Datong University under Grant No. 2014B06.

Conflicts of Interest: The authors declare no conflict of interest.

References

1. Shyllon, F. International Standards for Cultural Heritage: An African Perspective. *Int. Crimes* **2017**, *5*, 347–363.
2. Turkington, A.; Martin, E.; Viles, H.; Smith, B. Surface change and decay of sandstone samples exposed to a polluted urban atmosphere over a six-year period: Belfast, Northern Ireland. *Build. Environ.* **2003**, *38*, 1205–1216. [CrossRef]
3. Doehne, E.; Price, C.A. *Stone Conservation: An Overview of Current Research*, 2nd ed.; Getty Conservation Institute: Los Angeles, CA, USA, 2010; pp. 1–25.
4. Gericke, O.R. Determination of the Geometry of Hidden Defects by Ultrasonic Pulse Analysis Testing. *J. Acoust. Soc. Am.* **1963**, *35*, 364. [CrossRef]
5. Jiles, D. Review of magnetic methods for nondestructive evaluation (Part 2). *NDT E Int.* **1990**, *23*, 83–92. [CrossRef]
6. Márquez, E.; Ramirez-Malo, J.; Villares, P.; Jiménez-Garay, R.; Ewen, P.J.S.; Owen, A.E. Calculation of the thickness and optical constants of amorphous arsenic sulphide films from their transmission spectra. *J. Phys. D Appl. Phys.* **1992**, *25*, 535–541. [CrossRef]
7. Hellier, C.J. *Handbook of Nondestructive Evaluation*; Mcgraw-hill: New York, NY, USA, 2001.
8. Leona, M.; Casadio, F.; Bacci, M.; Picollo, M. Identification of the Pre-Columbian Pigment Mayablue on Works of Art by Noninvasive UV-Vis and Raman Spectroscopic Techniques. *J. Am. Inst. Conserv.* **2004**, *43*, 39–54. [CrossRef]
9. Anouncia, S.; Saravanan, R. Non-destructive testing using radiographic images? A survey. *Insight Non-Destructive Test. Cond. Monit.* **2006**, *48*, 592–597. [CrossRef]
10. Puryear, C.I.; Castagna, J.P. Layer-thickness determination and stratigraphic interpretation using spectral inversion: Theory and application. *Geophysics* **2008**, *73*, R37–R48. [CrossRef]
11. Duling, I.; Zimdars, D. Revealing hidden defects. *Nat. Photon* **2009**, *3*, 630–632. [CrossRef]
12. Lü, Q.; Tang, M.-J.; Cai, J.; Zhao, J.-W.; Vittayapadung, S. Vis/NIR hyperspectral imaging for detection of hidden bruises on kiwifruits. *Czech J. Food Sci.* **2011**, *29*, 595–602. [CrossRef]
13. Duvillaret, L.; Garet, F.; Coutaz, J.-L. Highly precise determination of optical constants and sample thickness in terahertz time-domain spectroscopy. *Appl. Opt.* **1999**, *38*, 409–415. [CrossRef]
14. Ospina-Borras, J.E.; Benitez-Restrepo, H.D.; Benitez-Restrepo, H.D. Non-Destructive Infrared Evaluation of Thermo-Physical Parameters in Bamboo Specimens. *Appl. Sci.* **2017**, *7*, 1253. [CrossRef]
15. Zhang, W.; Lei, Y. Progress in terahertz nondestructive testing. *Chin. J. Sci. Instrum.* **2008**, *29*, 1563–1568. [CrossRef]
16. Jolly, M.; Prabhakar, A.; Sturzu, B.; Hollstein, K.; Singh, R.; Thomas, S.; Foote, P.; Shaw, A. Review of Non-destructive Testing (NDT) Techniques and their Applicability to Thick Walled Composites. *Procedia CIRP* **2015**, *38*, 129–136. [CrossRef]
17. Park, S.-H.; Jang, J.-W.; Kim, H.-S. Non-destructive evaluation of the hidden voids in integrated circuit packages using terahertz time-domain spectroscopy. *J. Micromech. Microeng.* **2015**, *25*, 95007. [CrossRef]
18. Cheng, L.; Wang, L.; Mei, H.; Guan, Z.; Zhang, F. Research of nondestructive methods to test defects hidden within composite insulators based on THz time-domain spectroscopy technology. *IEEE Trans. Dielectr. Electr. Insul.* **2016**, *23*, 2126–2133. [CrossRef]
19. Oh, G.-H.; Jeong, J.-H.; Park, S.-H.; Kim, H.-S. Terahertz time-domain spectroscopy of weld line defects formed during an injection moulding process. *Compos. Sci. Technol.* **2018**, *157*, 67–77. [CrossRef]
20. Lewis, R. A review of terahertz detectors. *J. Phys. D Appl. Phys.* **2019**, *52*, 433001. [CrossRef]
21. Wang, Y.; Sun, Z.; Xu, D.; Wu, L.; Chang, J.; Tang, L.; Jiang, Z.; Jiang, B.; Wang, G.; Chen, T.; et al. A hybrid method based region of interest segmentation for continuous wave terahertz imaging. *J. Phys. D Appl. Phys.* **2019**, *53*, 095403. [CrossRef]
22. Akevren, S. Non-Destructive Examination of Stone Masonry Historic Structures Quantitative IR Thermography and Ultrasonic Testing. Master's Thesis, Middle East Technical University, Universiteler Mahallesi, Cankaya Ankara, Turkey, 2010.
23. Wang, Y.; Xia, Y.; Zhang, J.S.; Li, H.S.; Dai, S.B.; Tang, Z. Experimental Research about Weathering Resistance and Surface Deterioration of Two Kinds of Stone Cultural Relics. *Adv. Mater. Res.* **2011**, *250*, 65–69. [CrossRef]

24. Li, L.; Zhou, M.; Ren, J. Test of the adhesive thickness uniformity based on terahertz time-domain spectroscopy. *Laser Infrared.* **2014**, *44*, 801–804. [CrossRef]
25. Tanaka, S.; Shiraga, K.; Ogawa, Y.; Fujii, Y.; Okumura, S. Applicability of effective medium theory to wood density measurements using terahertz time-domain spectroscopy. *J. Wood Sci.* **2014**, *60*, 111–116. [CrossRef]
26. Zhang, Z.; Wang, K.; Lei, Y.; Zhang, Z.; Zhao, Y.; Li, C.; Gu, A.; Shi, N.; Zhao, K.; Zhan, H.; et al. Non-destructive detection of pigments in oil painting by using terahertz tomography. *Sci. China Ser. G Phys. Mech. Astron.* **2015**, *58*, 124202. [CrossRef]
27. Yang, Y.; Wu, T.V.; Sempey, A.; Pradère, C.; Sommier, A.; Batsale, J.-C. Combination of terahertz radiation method and thermal probe method for non-destructive thermal diagnosis of thick building walls. *Energy Build.* **2018**, *158*, 1328–1336. [CrossRef]
28. Swanepoel, R. Determination of the thickness and optical constants of amorphous silicon. *J. Phys. E Sci. Instrum.* **1983**, *16*, 1214–1222. [CrossRef]
29. Auston, D.H.; Cheung, K.P.; Valdmanis, J.A.; Kleinman, D.A. Cherenkov Radiation from Femtosecond Optical Pulses in Electro-Optic Media. *Phys. Rev. Lett.* **1984**, *53*, 1555–1558. [CrossRef]
30. Fattinger, C.; Grischkowsky, D. Terahertz beams. *Appl. Phys. Lett.* **1989**, *54*, 490–492. [CrossRef]

Article

Improved InGaAs and InGaAs/InAlAs Photoconductive Antennas Based on (111)-Oriented Substrates

Kirill Kuznetsov [1,*], Aleksey Klochkov [2,*], Andrey Leontyev [1], Evgeniy Klimov [2], Sergey Pushkarev [2], Galib Galiev [2] and Galiya Kitaeva [1]

1 Faculty of Physics, Lomonosov Moscow State University, Moscow 119991, Russia; aa.leontjev@physics.msu.ru (A.L.); gkitaeva@physics.msu.ru (G.K.)
2 V.G. Mokerov Institute of Ultra High Frequency Semiconductor Electronics of Russian Academy of Sciences, Moscow 117105, Russia; klimov_evgenyi@mail.ru (E.K.); s_s_e_r_p@mail.ru (S.P.); galiev_galib@mail.ru (G.G.)
* Correspondence: kirill-spdc@yandex.ru (K.K.); klochkov_alexey@mail.ru (A.K.)

Received: 28 February 2020; Accepted: 14 March 2020; Published: 17 March 2020

Abstract: The terahertz wave generation by spiral photoconductive antennas fabricated on low-temperature $In_{0.5}Ga_{0.5}As$ films and $In_{0.5}Ga_{0.5}As/In_{0.5}Al_{0.5}As$ superlattices is studied by the terahertz time-domain spectroscopy method. The structures were obtained by molecular beam epitaxy on GaAs and InP substrates with surface crystallographic orientations of (100) and (111)A. The pump-probe measurements in the transmission geometry and Hall effect measurements are used to characterize the properties of LT-InGaAs and LT-InGaAs/InAlAs structures. It is found that the terahertz radiation power is almost four times higher for LT-InGaAs samples with the (111)A substrate orientation as compared to (100). Adding of LT-InAlAs layers into the structure with (111)A substrate orientation results in two orders of magnitude increase of the structure resistivity. The possibility of creating LT-InGaAs/InAlAs-based photoconductive antennas with high dark resistance without compensating Be doping is demonstrated.

Keywords: terahertz wave generation; InGaAs; molecular beam epitaxy; time-domain spectroscopy; photoconductive antenna

1. Introduction

At the present time terahertz wave radiation is widely used in various application fields related to pharmacology, medicine, security systems, and data transmission. Therefore, a search is underway for the most effective methods for generating and detecting terahertz radiation. In this regard, photoconductive semiconductor antennas (PCAs) have proven themselves as flexible and effective THz devices for use in time-domain spectroscopy (TDS) and imaging systems. Due to the availability of cost-efficient and reliable components from the telecommunication industry, the state-of-the-art TDS systems often utilize optical excitation at 1550 nm [1,2] with Er^{3+} fiber laser femtosecond pulses. Recently [3,4], the $In_{0.5}Ga_{0.5}As$-based semiconductor structures have been investigated as photoconductive materials for THz PCAs due to an appropriate room-temperature optical absorption cut-off wavelength of 1.5 μm. Lately, a significant improvement in the performance of these devices was achieved, mainly with the help of nanotechnology tools such as plasmonic light concentrators, plasmonic contact electrodes, optical nanoantenna arrays, or optical nanocavities [5–7]. Nevertheless, the development of basic photoconductive materials for THz PCAs is still of current interest.

In order to generate fast transient current and to sample the THz pulse accurately in the time domain, THz PCAs require a photoconductor with high dark resistivity and a short carrier lifetime after optical excitation. Additionally, high electron mobility is necessary for THz PCA detectors. The main development challenge with InGaAs-based photoconductors is relatively small bandgap resulting in low breakdown field strength and large dark background conductivity. Several methods for producing ultrafast InGaAs photoconductive structures have been proposed: ion implantation with heavy ions followed by thermal annealing treatment [8,9], epitaxial doping of InGaAs with impurities producing deep electronic traps in the bandgap [10,11], and growth of specially designed heterostructures such as ErAs inclusions in InGaAlAs [12]. The low-temperature (LT) InGaAs-based epitaxial structures are widely used in commercial THz systems. Low substrate temperatures ($T_G \approx 200$ °C) lead to non-stoichiometric growth with the incorporation of excess arsenic in the crystal structure. The most common non-stoichiometry related point defect is arsenic antisite with concentration in the range of 10^{17}–10^{19} cm^{-3} depending on the substrate temperature and arsenic overpressure [13–15]. The fast capture of photogenerated electrons and recombination with holes through antisite centers results in sub-picosecond carrier lifetimes in LT-materials at optimized growth and annealing conditions [16]. LT-InGaAs shows a high room-temperature residual electron concentration of the order of 10^{17} cm^{-3} due to the thermal ionization of antisite defects. To increase the structure resistivity, LT-InAlAs layers are added to the photoconductive material. LT-InAlAs layers have a higher dark resistivity as compared to LT-InGaAs and exhibit deep trap states that are situated energetically below the antisite defect levels of adjacent InGaAs layers. With acceptor doping by beryllium (Be) atoms, LT-InGaAs/InAlAs superlattices demonstrate both low residual electron concentration and short carrier lifetimes in the sub-picosecond range [17–19].

It was shown recently [20–22] that LT-InGaAs and LT-GaAs structures, fabricated on (n11)A-oriented GaAs or InP substrates with or without PCA electrodes, under pulsed laser excitation, can generate THz radiation with higher power as compared to structures obtained on conventional (100) substrates. It was argued that substrate orientation can influence the concentration and the type of defects in low-temperature grown films. The purpose of the present work is further clarification of the nature of this effect by investigating the carrier dynamics in LT-InGaAs-based epitaxial structures grown on (111)A GaAs and InP substrates. Among the structures, for the first time we study the PCA based on LT-InGaAs/InAlAs superlattice fabricated on InP (111)A-oriented substrate.

2. Materials and Methods

Figure 1 shows the schematic of photoconductive terahertz source, which consists of a spiral antenna fabricated on an epitaxial heterostructure. As the photoconductive heterostructures we investigate undoped $In_{0.53}Ga_{0.47}As$ layers with metamorphic buffer on GaAs substrates [20] and an undoped lattice-matched $In_{0.53}Ga_{0.47}As/In_{0.52}Al_{0.48}As$ superlattice on InP substrate. The $In_{0.53}Ga_{0.47}As$ layers serve as the photo-absorbing active regions. The mole fraction of indium is chosen as 0.53 to provide an optical absorption cut-off wavelength close to the operating wavelength of 1.5 μm. The room temperature absorption coefficient for $In_{0.53}Ga_{0.47}As$ layers lattice matched to InP substrates is 8000–12000 cm^{-1} in this wavelength range [23,24]. The thickness of LT-$In_{0.53}Ga_{0.47}As$ layers is chosen as 660 nm to be of the order of the excitation light penetration depth. Also, the $In_{0.53}Ga_{0.47}As$ layer is sufficiently thin to minimize the dark conductivity. The $In_{0.52}Al_{0.48}As$ layers in superlattice structures are used to introduce deep defect states and to reduce residual carrier concentration.

The active layers of the photoconductive structures are grown by molecular-beam epitaxy (MBE) at a low growth temperature of 200 °C. Two types of structure layer designs on different substrate materials and orientations are investigated. The 660-nm-thick $In_{0.53}Ga_{0.47}As$ layers are grown on GaAs substrates with crystallographic orientations (100) and (111)A. The step-graded metamorphic buffer $In_xGa_{1-x}As$ is used to accommodate the lattice mismatch between the GaAs substrate and the photoconductive layer. The metamorphic buffer consists of eleven 60-nm-thick $In_xGa_{1-x}As$ layers with indium composition increment of $\Delta x = 0.05$. The substrate temperature during metamorphic buffer growth is stepwise decreased from 390 °C for $In_{0.05}Ga_{0.95}As$ to 320 °C for the topmost $In_{0.55}Ga_{0.45}As$ layer. The beam equivalent pressure (BEP) of arsenic molecules As_4 is 2.0×10^{-5} Torr during growth.

The III/V BEP ratio is 29. The 100-period $In_{0.53}Ga_{0.47}As$ (12 nm)/$In_{0.52}Al_{0.48}As$ (8 nm) superlattice is epitaxially grown on the InP (111)A substrate. The BEP of As_4 is 1.2×10^{-5} Torr and III/V BEP ratio is 30. The samples were in-situ annealed in the growth chamber for 30 min at 590 °C. On the surface of heterostructures the spiral antennas with 320 nm-thick Ni/Ge/Au/Ni/Au annealed ohmic electrodes are deposited using standard photolithography. The central photoconductive gap is 20 μm. The Hall mobility and density of charge carriers were measured using the Van der Pauw method at room temperature.

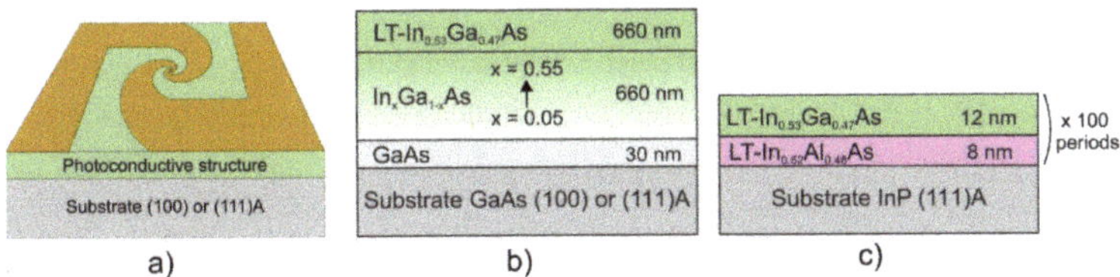

Figure 1. Schematic of the investigated photoconductive semiconductor antenna (PCA) THz emitters: (**a**) topology of spiral antenna, (**b**) LT-InGaAs/GaAs (100) and (111)A heterostructures, (**c**) LT-InGaAs/InAlAs/InP (111)A heterostructure.

One of the most common methods for studying the dynamics of relaxation of charge carriers in semiconductor structures is based on the pump-probe technique. This method has sufficient resolution to observe the dynamics of relaxation of charge carriers on a sub-picosecond time scale [17].

Figure 2 shows our experimental setup for frequency-degenerate pump-probe measurements with time resolution. We measure the temporal behavior of the differential transmission (DT) of the samples induced by pump pulses. The optical radiation source (Las) is a femtosecond Er^{3+} laser (wavelength 1550 nm, pulse duration 100 fs, pulse repetition rate 70 MHz). The initial laser beam was divided into probe and pump beams. The pump and probe beams were separated on a Glan polarization prism (G) and had orthogonal polarizations. The power ratio between the beams was regulated using a λ/2 plate in front of the Glan prism (G). A pump beam was focused on the sample; it modulated transmittance of the sample due to interband transitions of electrons in a semiconductor film. The time shift between the pulses of pump and probe beams was realized using the corner reflector (C) mounted on the delay line. The shift was changed with the time step Δt = 0.1 ps. The modulation frequency of the pump radiation chopper (Ch) was chosen equal to 2.3 kHz. The mean pump beam power was 40 mW, the probe beam power was 5 mW. The focal length of the lens (L), which focused the both beams on the sample (S), was f = 1.5 cm. An InGaAs-detector (D) by Thorlabs was used to measure the probe differential transmittance. The signal from the detector was recorded using a Lock-in Amplifier (Amp).

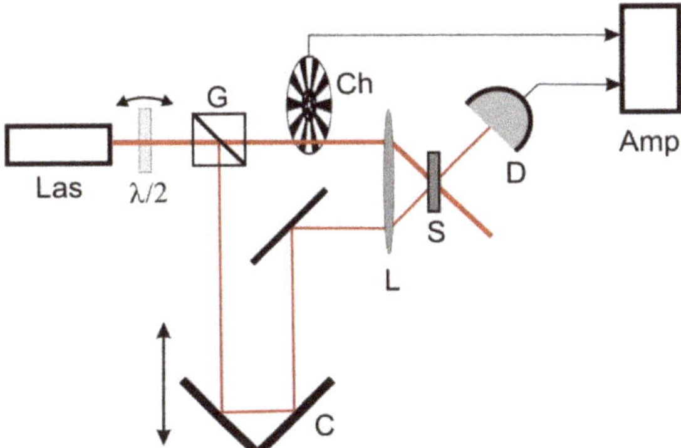

Figure 2. Schematics of the pump-probe setup for frequency-degenerate differential transmission (DT) measurements.

Figure 3 shows the experimental setup for generation and detection of terahertz radiation. The source of optical radiation was the same femtosecond Er^{3+} laser as in pump-probe experiments. After passing through the beam splitter, most of the radiation was directed to a generating photoconductive antenna, and the rest served to illuminate the detector. The pump beam (with a mean power up to 30 mW) was focused by a lens with a focal length of 5 mm onto the studied antennas. For collimation of the generated THz radiation, a matching silicon lens was placed on the output surface of the antenna-generator (PCA1). The refractive index of high-resistive silicon is n_{Si} = 3.14 in the terahertz range. Four parabolic mirrors were used to collect THz radiation and to focus it further on the silicon lens of the antenna-detector (PCA2). The commercial antenna-detector (Menlo Systems) with a symmetrical dipole antenna located on its back surface was used to detect the THz radiation. An instantaneous action of the THz field on an antenna-detector is equivalent to inducing some difference in potentials of its electrodes. However, the corresponding current arises through the antenna only upon irradiation with an optical pulse, which starts the process of generating free charge carriers in a semiconductor wafer. In order to register the temporal dynamics of the THz field, it is necessary to introduce a controlled delay between the pump beam and the probe beam. This is done using a mechanical delay line with a corner reflector installed. Thus, by changing the arrival time of the probe pulse (using a moving delay line), one can measure various instantaneous values of the terahertz field and, thereby, determine its temporal shape of the THz pulse. Detection of the current was performed using a Lock-in Amplifier. For synchronous detection of radiation, a radiation chopper was placed in the optical path, which modulated the laser beam at a frequency f = 2.3 kHz.

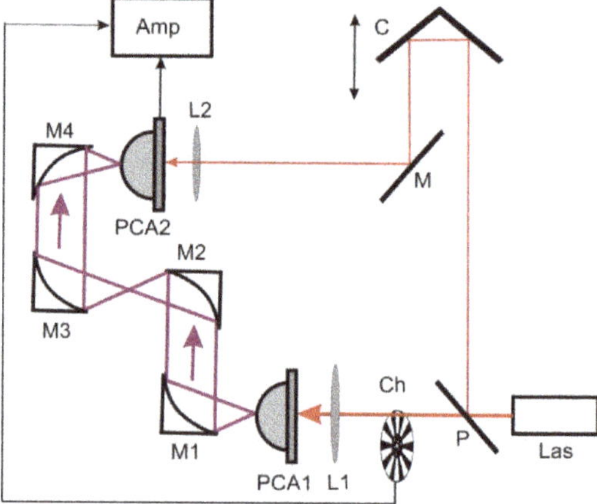

Figure 3. Schematics of the terahertz time-domain spectroscopy setup.

3. Experimental Results

The room-temperature Hall data is presented in Table 1. The electron-type conductivity of the investigated samples is much greater than that of the reference samples of thick undoped $In_{0.53}Ga_{0.47}As$ films lattice-matched to InP (100) substrates obtained at normal growth conditions. The typical unintentional background doping of InGaAs layers in our MBE system is (1–2) $\cdots$ 10^{15} cm^{-3}. The increase of the electron concentration at low growth temperature of 200 °C corresponds to the incorporation of excess As atoms into the InGaAs lattice. The thermal ionization of antisite As defects leads to the free electron volume concentration of the order of 10^{17} cm^{-3}. The electron sheet concentration and mobility depend on the substrate orientation and on the heterostructure design. From the comparison of the LT-InGaAs samples it follows that MBE growth on the (100) and (111)A GaAs substrates at identical growth conditions leads to formation of InGaAs layers with different concentration of point defects. It can be concluded that the antisite defect density is greater in the case of the sample grown on (111)A GaAs substrate. Introduction of InAlAs barrier layers into InGaAs/InAlAs superlattice results in substantial decrease of the sheet electron concentration by an order of magnitude as compared to thick InGaAs layers. The addition of InAlAs layers also leads to reduction of mobility μ. As a result, the dark resistivity of LT-InGaAs/InAlAs (111)A sample is 60 and 130 times greater than that of the LT-InGaAs/GaAs (111)A and (100) samples, correspondingly. The effect of InAlAs layers on the electron transport is associated with the carrier capture by deep traps into InAlAs barriers and with the scattering.

Table 1. Electronic mobility μ and sheet concentration n_S in photoconductive heterostructures.

Sample	n_S (10^{12} cm^{-2})	μ (cm^2/Vs)
LT-InGaAs/GaAs (100)	11.9	380
LT-InGaAs/GaAs (111)A	20.1	110
LT-InGaAs/InAlAs/InP (111)A	1.2	30

The pump-probe measurements were performed for LT-InGaAs samples grown on GaAs substrates with the surface crystallographic orientations of (100) and (111)A and for the LT-InGaAs/InAlAs superlattice grown on an InP substrate with (111)A orientation. Figure 4 shows graphs of the temporal dependences of the normalized transmittance for the studied samples. It should be noted that on some graphs, in the falling part of the functional dependences, there are regions of local growth of the normalized reflection coefficient function. These areas appear due to re-reflection of the signal from the back side of the samples. Points in these areas (dashed rectangles) were cut out of the dependencies for correct subsequent approximation.

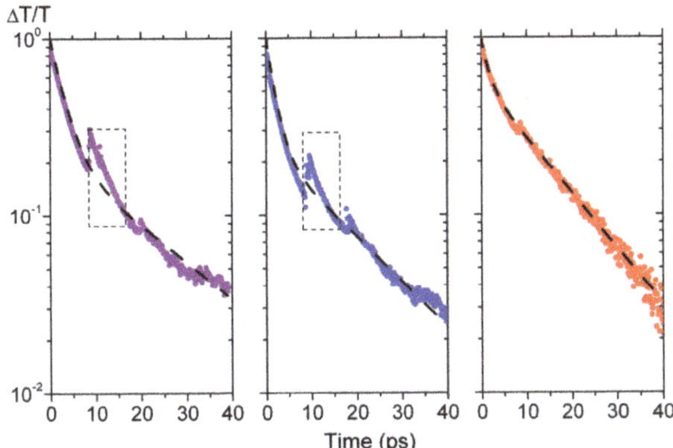

Figure 4. Normalized transmission changes detected for: (left) LT-InGaAs/GaAs(100), (middle) LT-InGaAs/GaAs (111)A, (right) LT-InGaAs/InAlAs/InP (111)A.

Figure 5 shows the temporal profiles of terahertz radiation detected from the studied samples. The measurements were carried out at the same input sensitivity of the Lock-in Amplifier. The magnitude of the applied bias voltage in the case of LT-InGaAs samples was 3.7 V. It can be seen that the signal in the sample grown on a (111)A GaAs substrate is 1.9 times higher than that from the sample grown on (100) GaAs substrate. For the LT-InGaAs / InAlAs (InP) (111)A sample, the applied bias voltage was higher almost by an order of magnitude, equal to 25 V. It can be seen that, mostly due to the higher bias voltage, the signal from the InGaAs / InAlAs-based structure is 5-6 times higher. Applying of higher bias voltage turned out to be possible (without breakdown of the sample) due to significantly higher resistance of this antenna and its lower dark current.

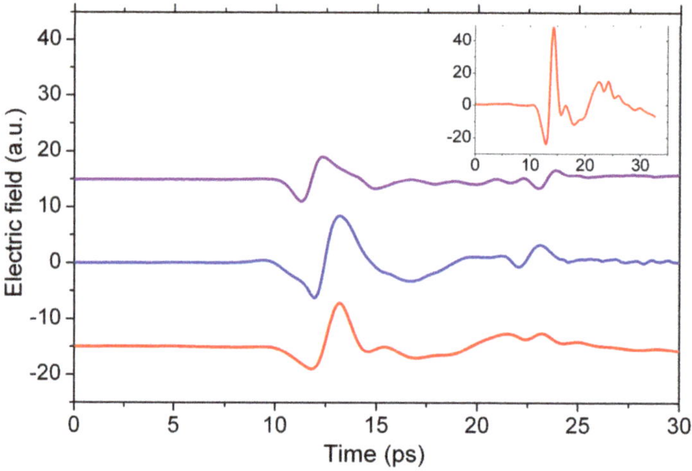

Figure 5. Detected time-domain signal traces: violet line LT-InGaAs/GaAs(100) U_b = 3.7 V, blue line LT-InGaAs/GaAs (111)A U_b = 3.7 V, red line LT-InGaAs/InAlAs/InP(111)A U_b = 3.7 V, (inset) LT-InGaAs/InAlAs/InP(111)A U_b = 25 V.

Figure 6 shows the normalized spectra obtained by the fast Fourier transform (FFT) processing of the temporal waveforms presented in Figure 5. It can be seen that the maximum of the power spectral distribution is located at the frequency of 0.2 THz, and the total width of each spectrum is about 2 THz. We have not found significant differences in the spectra of terahertz wave radiation from the studied antennas. Apparently, this is due to the limiting effect of the antenna-detector with an upper frequency of the detection band about 2 THz.

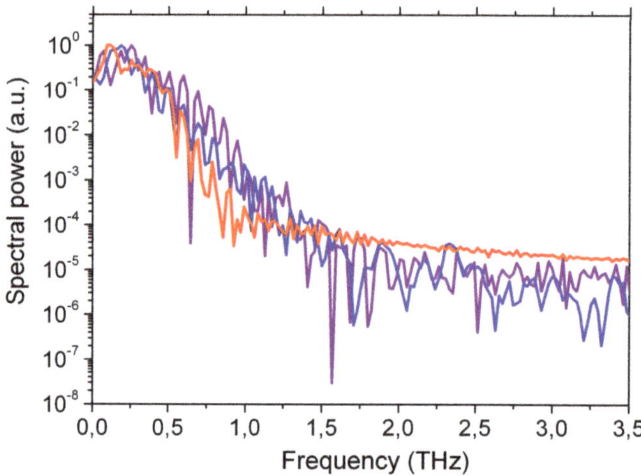

Figure 6. FFT spectra of antennas: violet line LT-InGaAs/GaAs(100), blue line LT-InGaAs/GaAs (111)A, red line LT-InGaAs/InAlAs/InP (111)A.

4. Discussion

4.1. Pump-Probe Results.

Figure 4 shows the experimental temporal dependences of the falling part of the normalized transmittance. Two separated regions (up to 10 ps, after 10 ps) with different slopes of the curve could be clearly distinguished on each of the presented dependencies. Since the curves are plotted on a logarithmic scale, we can assume these regions as two exponential contributions with different characteristic relaxation times. Dependencies are well approximated by a two-exponential decay model. The interpolation was carried out using the least squares method by the following expression:

$$\frac{\Delta T(t)}{T_0} = Ae^{-t/\tau_1} + Be^{-t/\tau_2} \quad (1)$$

Here ΔT is a DT, T_0 is a maximal difference transmission, τ_1 and τ_2 are the relaxation times, A and B are the fitting constants. Fitting results are presented in Table 2.

Table 2. Characteristic relaxation times.

Sample	τ_1 (ps)	τ_2 (ps)
LT-InGaAs/GaAs(100)	3.0 ± 0.1	21.0 ± 0.2
LT-InGaAs/GaAs(111)A	1.9 ± 0.1	17.0 ± 0.3
LT-InGaAs/InAlAs/InP(111)A	1.7 ± 0.1	14.0 ± 0.2

Most likely, the characteristic time τ_1 is the time of an electron capture from the conduction band by an anti-structural defect As_{Ga}^+. The obtained values of τ_1 are in a good agreement with the data of experiments performed earlier by other scientific groups [2,16,25]. The time interval τ_2 seems to be the recombination time of holes and electrons captured by traps. Its characteristic scale is of the order of tens of picoseconds, which, in order of magnitude, agrees well with the previous experimental values [17]. The origin of the processes that occur with photo-excited electrons at the short capture time τ_1 is connected with the excess of arsenic atoms in the InGaAs crystal structure leading to formation of the special-type defects.

As it is known from the theory by Shockley and Read [26], the carrier capture time by traps (τ_1) is determined by the following expression:

$$\tau_1 = (N_{As+} v_{th} \sigma_{As+})^{-1} \quad (2)$$

Here, N_{As+} is concentration of As+ traps in the crystal, σ_{As+} is cross-section of the electron capture by traps, and v_{th} is the thermal velocity of electrons. Based on the obtained data, we can conclude that the number of As_{Ga}^+ defects in the LT-InGaAs/GaAs (111)A and LT-InGaAs/InAlAs/InP (111)A samples is almost two times higher than that in LT-InGaAs / GaAs (100).

4.2. Temporal and Spectral Dependences of the Generated THz Field.

Terahertz-wave electric field should be approximately proportional to the derivative of the current density. Indeed, in the simple case of a weak pump, neglecting screening effects, quasi-static and near-field terms in the expression for the electric field, one can obtain the terahertz-wave electric field proportional to the derivative of the concentration of the free electrons [27]. In Figure 5, we see that the terahertz signal amplitude from the sample LT-InGaAs/GaAs (111)A is almost two times higher than that from the sample LT-InGaAs/GaAs (100). This can be explained under the assumption that concentration of charged arsenic traps with (111)A substrate orientation is greater than that with (100) orientation. This assumption is in good agreement with the results of our measurements of characteristic electron capture times. Slower dynamics of charge carriers here could be also explained by decrease in the concentration of active defects As_{Ga}+.

It is usually assumed that fast excitation of charge carriers in InGaAs / InAlAs heterostructures occurs in InGaAs layers, whereas the diffused charge carriers are captured in InAlAs layers [28,29]. Since it was shown that the most efficient terahertz radiation was generated in samples grown on substrates with an orientation different from the traditional (100) [20–22], to obtain the most efficient generation, the InGaAs / InAlAs heterostructure was grown straight on InP (111)A substrate. InAlAs layers significantly increase the resulting structure resistance. The resistance of the InGaAs / InAlAs sample on the InP (111)A substrate was 12.8 kΩ, which is significantly higher than the resistance of 30 Ω for all InGaAs samples on GaAs substrates. This made it possible to increase the bias voltage from 3.7 to 25 V without damaging the antenna and to raise the terahertz radiation power by almost 30 times. Since the reflections are present in the waveforms in Figure 5, the resulting spectra in Figure 6 are strongly modulated with a frequency inversely proportional to the delay between the main and reflected pulses. However, the main features of the antennas are clearly seen. The spectrum from the antenna on the LT-InGaAs/InAlAs heterostructure looks more pronounced and less "noisy" at high frequencies above 2 THz than from the LT-InGaAs antennas. Although the total spectral widths of all antennas on the logarithmic scale practically coincide in Figure 6, the total spectrum of the antenna on the heterostructure InGaAs/InAlAs is slightly wider than that of InGaAs. This issue needs a more detailed study using a wide range of radiation detectors to exclude the possible influence of the receiver on the measured spectrum.

For future research directions, we propose using atypically oriented substrates for the manufacture of antennas in order to obtain faster carrier dynamics, larger signals, and wider terahertz radiation spectra.

5. Conclusions

The characteristics of terahertz wave radiation from photoconductive antennas based on epitaxial films of low-temperature grown InGaAs with orientations of the crystallographic axes of the GaAs substrate (111)A and (100) were studied. It was found that the terahertz radiation power is almost four times higher for samples with the (111)A substrate orientation. The observed increase in the radiation power is associated with an increase in the number of anti-structural defects. THz radiation generated in the LT-InGaAs / InAlAs / InP (111)A heterostructure was 25 times higher than in the LT-InGaAs / GaAs (111)A antenna. The characteristic relaxation times of charge carriers were measured in LT-InGaAs samples on GaAs substrates with (111)A and (100) orientations, as well as in the LT-InGaAs/InAlAs/InP (111)A heterostructure. Obtained values are consistent with previously published data and qualitatively confirm the proposed explanation of the advantages of the non-standard orientation of the antenna substrate.

Author Contributions: Conceptualization, G.G.; MBE growth, E.K. and A.K.; Hall effect measurements, A.K.; PCA antenna, S.P.; pump-probe measurements, A.L. and K.K.; TDS measurements, A.L. and K.K.; measurement analysis, G.K.; data fitting, A.L.; writing and editing, K.K. and A.K. All authors have read and agreed to the published version of the manuscript.

Funding: This research was funded by RFBR, grants numbers 18-29-20101, 18-32-20207, 19-02-00598.

Acknowledgments: The authors thank D.S. Ponomarev for some helpful discussions, D.V. Lopaev for the help with experiments and I.S. Vasil'evskii, A.N. Vinichenko for help with MBE.

Conflicts of Interest: The authors declare no conflict of interest.

References

1. Vieweg, N.; Rettich, F.; Deninger, A.; Roehle, H.; Dietz, R.; Göbel, T.; Schell, M. Terahertz-time domain spectrometer with 90 dB peak dynamic range. *J. Infrared Milli Terahz Waves* **2014**, *35*, 823–832. [CrossRef]
2. Globisch, B.; Dietz, R.J.B.; Kohlhaas, R.B.; Göbel, T.; Schell, M.; Alcer, D.; Semtsiv, M.; Masselink, W.T. Iron doped InGaAs: Competitive THz emitters and detectors fabricated from the same photoconductor. *J. Appl. Phys.* **2017**, *121*, 053102. [CrossRef]

3. Castro-Camus, E.; Alfaro, M. Photoconductive devices for terahertz pulsed spectroscopy: A review. *Photon. Res.* **2016**, *4*, A36. [CrossRef]
4. Burford, N.M.; El-Shenawee, M.O. Review of terahertz photoconductive antenna technology. *Opt. Eng.* **2017**, *56*, 010901. [CrossRef]
5. Yardimci, N.T.; Jarrahi, M. Nanostructure-Enhanced Photoconductive Terahertz Emission and Detection. *Small J.* **2018**, *14*, 1802437. [CrossRef]
6. Lepeshov, S.; Gorodetsky, A.; Krasnok, A.; Rafailov, E.; Belov, P. Enhancement of terahertz photoconductive antenna operation by optical nanoantennas: Enhancement of terahertz photoconductive antenna operation by optical nanoantennas. *Laser Photonics Rev.* **2017**, *11*, 1600199. [CrossRef]
7. Yachmenev, A.E.; Lavrukhin, D.V.; Glinskiy, I.A.; Zenchenko, N.V.; Goncharov, Y.G.; Spektor, I.E.; Khabibullin, R.A.; Otsuji, T.; Ponomarev, D.S. Metallic and dielectric metasurfaces in photoconductive terahertz devices: A review. *Opt. Eng.* **2019**, *59*, 061608. [CrossRef]
8. Petrov, B.; Fekecs, A.; Sarra-Bournet, C.; Ares, R.; Morris, D. Terahertz Emitters and Detectors Made on High-Resistivity InGaAsP:Fe Photoconductors. *IEEE Trans. THz Sci. Technol.* **2016**, *6*, 1–7. [CrossRef]
9. Suzuki, M.; Tonouchi, M. Fe-implanted InGaAs terahertz emitters for 1.56μm wavelength excitation. *Appl. Phys. Lett.* **2005**, *86*, 051104. [CrossRef]
10. Kohlhaas, R.B.; Breuer, S.; Nellen, S.; Liebermeister, L.; Schell, M.; Semtsiv, M.P.; Masselink, W.T.; Globisch, B. Photoconductive terahertz detectors with 105 dB peak dynamic range made of rhodium doped InGaAs. *Appl. Phys. Lett.* **2019**, *114*, 221103. [CrossRef]
11. Yardimci, N.T.; Lu, H.; Jarrahi, M. High power telecommunication-compatible photoconductive terahertz emitters based on plasmonic nano-antenna arrays. *Appl. Phys. Lett.* **2016**, *109*, 191103. [CrossRef] [PubMed]
12. Nandi, U.; Norman, J.C.; Gossard, A.C.; Lu, H.; Preu, S. 1550-nm Driven ErAs:In(Al)GaAs Photoconductor-Based Terahertz Time Domain System with 6.5 THz Bandwidth. *J. Infrared Milli Terahz Waves* **2018**, *39*, 340–348. [CrossRef]
13. Künzel, H.; Böttcher, J.; Gibis, R.; Urmann, G. Material properties of GaInAs grown on InP by low-temperature molecular beam epitaxy. *Appl. Phys. Lett.* **1992**, *61*, 1347–1349. [CrossRef]
14. Metzger, R.A. Structural and electrical properties of low temperature GaInAs. *J. Vac. Sci. Technol. B* **1993**, *11*, 798. [CrossRef]
15. Grandidier, B.; Chen, H.; Feenstra, R.M.; McInturff, D.T.; Juodawlkis, P.W.; Ralph, S.E. Scanning tunneling microscopy and spectroscopy of arsenic antisites in low temperature grown InGaAs. *Appl. Phys. Lett.* **1999**, *74*, 1439–1441. [CrossRef]
16. Baker, C.; Gregory, I.S.; Tribe, W.R.; Bradley, I.V.; Evans, M.J.; Linfield, E.H.; Missous, M. Highly resistive annealed low-temperature-grown InGaAs with sub-500fs carrier lifetimes. *Appl. Phys. Lett.* **2004**, *85*, 4965–4967. [CrossRef]
17. Dietz, R.J.B.; Globisch, B.; Roehle, H.; Stanze, D.; Göbel, T.; Schell, M. Influence and adjustment of carrier lifetimes in InGaAs/InAlAs photoconductive pulsed terahertz detectors: 6 THz bandwidth and 90dB dynamic range. *Opt. Express* **2014**, *22*, 19411. [CrossRef]
18. Ponomarev, D.S.; Gorodetsky, A.; Yachmenev, A.E.; Pushkarev, S.S.; Khabibullin, R.A.; Grekhov, M.M.; Zaytsev, K.I.; Khusyainov, D.I.; Buryakov, A.M.; Mishina, E.D. Enhanced terahertz emission from strain-induced InGaAs/InAlAs superlattices. *J. Appl. Phys.* **2019**, *125*, 151605. [CrossRef]
19. Kuenzel, H.; Biermann, K.; Nickel, D.; Elsaesser, T. Low-temperature MBE growth and characteristics of InP-based AlInAs/GaInAs MQW structures. *J. Cryst. Growth* **2001**, *227–228*, 284–288. [CrossRef]
20. Kuznetsov, K.A.; Galiev, G.B.; Kitaeva, G.K.; Kornienko, V.V.; Klimov, E.A.; Klochkov, A.N.; Leontyev, A.A.; Pushkarev, S.S.; Malrsev, P.P. Photoconductive antennas based on epitaxial films $In_{0.5}Ga_{0.5}As$ on GaAs (111)A and (100)A substrates with a metamorphic buffer. *Laser Phys. Lett.* **2018**, *15*, 076201. [CrossRef]
21. Galiev, G.B.; Grekhov, M.M.; Kitaeva, G.K.; Klimov, E.A.; Klochkov, A.N.; Kolentsova, O.S.; Kornienko, V.V.; Kuznetsov, K.A.; Maltsev, P.P.; Pushkarev, S.S. Terahertz-radiation generation in low-temperature InGaAs epitaxial films on (100) and (411) InP substrates. *Semiconductors* **2017**, *51*, 310–317. [CrossRef]
22. Galiev, G.B.; Pushkarev, S.S.; Buriakov, A.M.; Bilyk, V.R.; Mishina, E.D.; Klimov, E.A.; Vasil'evskii, I.S.; Maltsev, P.P. Terahertz-radiation generation and detection in low-temperature-grown GaAs epitaxial films on GaAs (100) and (111)A substrates. *Semiconductors* **2017**, *51*, 503–508. [CrossRef]
23. Bacher, F.R.; Blakemore, J.S.; Ebner, J.T.; Arthur, J.R. Optical-absorption coefficient of InGaAs/InP. *Phys. Rev. B* **1988**, *37*, 2551–2557. [CrossRef] [PubMed]

24. Zielinski, E.; Schweizer, H.; Streubel, K.; Eisele, H.; Weimann, G. Excitonic transitions and exciton damping processes in InGaAs/InP. *J. Appl. Phys.* **1986**, *59*, 2196–2204. [CrossRef]
25. Globisch, B.; Dietz, R.J.B.; Stanze, D.; Gobel, T.; Schell, M. Carrier dynamics in Beryllium doped low-temperature-grown InGaAs/InAlAs. *Appl. Phys. Lett.* **2014**, *104*, 172103-1-4. [CrossRef]
26. Shockley, W.; Read, W.T. Statistics of the Recombinations of Holes and Electrons. *Phys. Rev.* **1952**, *87*, 835–842. [CrossRef]
27. Auston, D.H.; Cheung, K.P.; Smith, P.R. Picosecond photoconducting Hertzian dipoles. *Phys. Lett.* **1984**, *45*, 284–286. [CrossRef]
28. Sartorius, B.; Roehle, H.; Kunzel, H.; Boettcher, J.; Schlak, M.; Stanze, D.; Venghaus, H.; Schell, M. All-fiber terahertz time-domain spectrometer operating at 1.5 μm telecom wavelengths. *Opt. Express* **2008**, *16*, 9565–9570. [CrossRef]
29. Dietz, R.J.B.; Globisch, B.; Gerhard, M.; Velauthapillai, A.; Stanze, D.; Roehle, H.; Koch, M.; Gobel, T.; Schell, M. 64 μW pulsed terahertz emission from growth optimized InGaAs/InAlAs heterostructures with separated photoconductive and trapping regions. *Appl. Phys. Lett.* **2013**, *103*, 061103-1-4. [CrossRef]

Article

Design and Measurement of a 0.67 THz Biased Sub-Harmonic Mixer

Guangyu Ji [1,2], Dehai Zhang [1,*], Jin Meng [1], Siyu Liu [1,2] and Changfei Yao [3]

1 CAS Key Laboratory of Microwave Remote Sensing, National Space Science Center, Chinese Academy of Sciences, Beijing 100190, China; guangyuji1@163.com (G.J.); mengjin@mirslab.cn (J.M.); liusiyu16@mails.ucas.edu.cn (S.L.)
2 University of Chinese Academy of Sciences, Beijing 100049, China
3 School of Electronic and Information Engineering, Nanjing University of Information Science and Technology, Nanjing 210044, China; yaocf1982@163.com
* Correspondence: zhangdehai@mirslab.cn; Tel.: +86-010-6258-6483

Received: 20 December 2019; Accepted: 9 January 2020; Published: 15 January 2020

Abstract: To effectively reduce the requirement of Local Oscillator (LO) power, this paper presents the design and measurement of a biased sub-harmonic mixer working at the center frequency of 0.67 THz in hybrid integration. Two discrete Schottky diodes were placed across the LO waveguide in anti-series configuration on a 50 μm thick quartz-glass substrate, and chip capacitors were not required. At the driven of 3 mW@335 GHz and 0.35 V, the mixer had a minimum measured Signal Side-Band (SSB) conversion loss of 15.3 dB at the frequency of 667 GHz. The typical conversion loss is 18.2 dB in the band of 650 GHz to 690 GHz.

Keywords: bias; sub-harmonic mixer; anti-series; Schottky diode; conversion loss

1. Introduction

Terahertz usually refers to the frequency band between 0.1 THz to 10 THz. In recent years, there is an urgent demand for receivers operating at terahertz frequency in radio astronomy, planetary exploration, and atmospheric remote sensing [1]. As is well known, it is difficult to produce high Local Oscillator (LO) power in sub-millimeter and Terahertz frequency for lacking power amplifiers [2,3]. Harmonic mixers are widely used in terahertz heterodyne receivers due to the advantage of reducing the LO frequency. Sub-harmonic mixing and fourth harmonic mixing are the most-utilized mixing methods, and conversion loss increases with the number of mixing times [4].

At the frequency below 600 GHz, unbiased sub-harmonic mixers can achieve good noise performance. In Reference [5], sub-harmonic mixers operating at 183 GHz and 366 GHz are designed based on 3.7 μm thick GaAs membrane. The Double Side-Band (DSB) conversion loss and noise temperature of the 183 GHz sub-harmonic mixer are 4.9 dB and 608 K respectively, and the 366 GHz mixer is 6.9 dB and 1220 K. In Reference [6], a sub-harmonic mixer is designed working at 190–240 GHz using a discrete Schottky diode. The DSB noise temperature is lower than 1500 K and the DSB conversion loss is less than 10 dB at the frequency band. In addition, some similar unbiased sub-harmonic mixer designs can be found in Reference [7–10].

The output power of the LO sources decreases with increasing frequency, especially when the operating frequency reaches 0.6 THz and above [11,12]. So, it is urgent to design biased sub-harmonic mixers to reduce the requirement for LO power. At present, biased sub-harmonic mixers working at 585 GHz [13], 874 GHz [14,15], 1.2 THz [16,17], and 1.2 THz [18] are designed and reported based on advanced GaAs membrane film process or frameless architecture. The biased mixers mentioned above are all based on monolithic integration technology, where the on-chip capacitor is required. But, the thickness of the GaAs substrate is less than 5 μm, which is easy to bend and expensive.

It is valuable to research the biased sub-harmonic mixer in hybrid integration to solve practical engineering problems of low LO power. Because of the big size and poor performance of discrete chip capacitors in terahertz frequency, it is unable to achieve good performance to mixers in hybrid integration of anti-parallel configuration. In this paper, the biased hybrid integration scheme is adopted. The scheme is theoretically derived for the first time and compared with the anti-parallel structure to analyze the advantages and disadvantages. Two discrete Schottky diodes are biased by DC voltage and placed across the LO waveguide in anti-series configuration without chip capacitors. The mode of the LO and RF signals are orthogonal in mixing, so the LO and RF ports are highly isolated. The bias voltage is feed from the IF port and separated by a bias-T.

Section 2 illustrates the theoretical analysis and comparison of different mixing topologies. Section 3 presents the detailed architecture, simulation process, and simulation results of the 0.67 THz biased sub-harmonic mixer. The measurement platform and results of the mixer are depicted in Section 3 at the same time. Section 4 gives the discussion and comparison in simulation, measurement results. Finally, the conclusion is presented in Section 5.

2. Comparison and Analysis of Different Mixing Topologies

Three different mixing topologies are shown in Figure 1. Figure 1a,b presents two anti-parallel circuit topologies that are commonly used in terahertz sub-harmonic mixers, including biased and unbiased mixing. The currents direction of the primary, secondary and tertiary harmonics of the LO and RF signals of the two diodes are illustrated in Figure 1. The total mixing current contains frequency terms $f = mf_{RF} \pm nf_{LO}$ listed in Table 1, where m and n is integer. To sub-harmonic mixers, the IF signal is the only concerned frequency which can be expressed as $f_{IF} = \left|f_{RF} - 2f_{LO}\right|$.

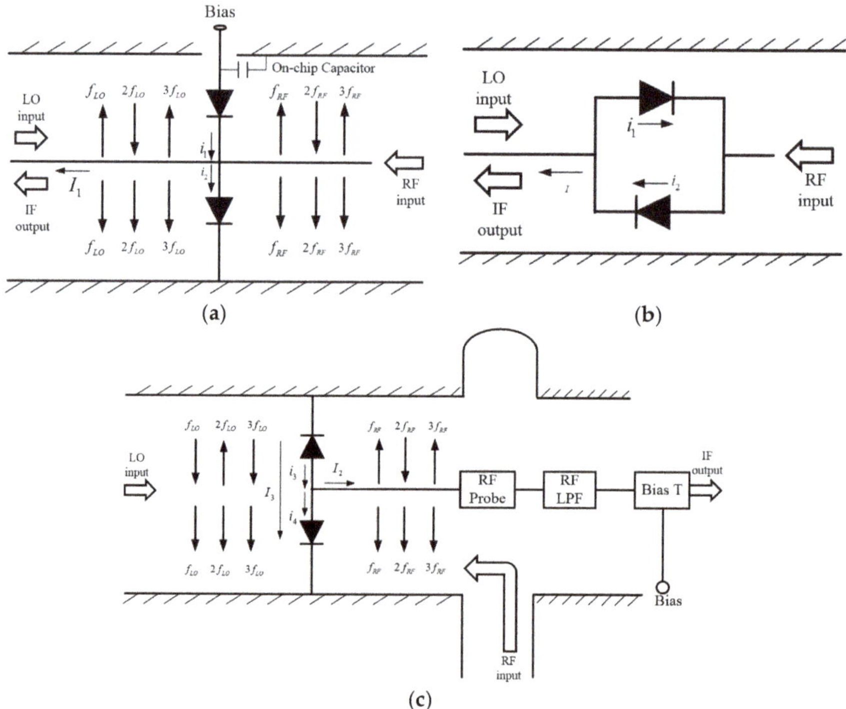

Figure 1. (**a**) Topology of biased anti-parallel; (**b**) topology of unbiased anti-parallel; (**c**) the topology of biased anti-series.

Table 1. Mixing products of anti-parallel and anti-series diodes configuration.

	Output Signal (f=mf_{RF}±nf_{LO})			
	m:odd n:odd	m:even n:even	m:odd n:even	m:even n:odd
Anti-Parallel I_1	×	×	√	√
Anti-Series $I_2(I_3)$	× (√)	√ (×)	√ (×)	× (√)

Sub-harmonic mixers using the topology shown in Figure 1a typically are based on monolithic integration due to the one on-chip capacitor required. It has the same equivalent circuit with the topology in Figure 1b, which is the anti-parallel mixing configuration. The RF and LO signals are feed to diodes in quasi-Transverse Electromagnetic (TEM) mode. In Figure 1a,b, i_1 and i_2 have the same frequency components. The only output current I can be expressed as $I_1 = i_2 - i_1$. It means that there is output only when i_1 and i_2 are in the opposite phase. The currents through the diode junctions can be written as

$$\begin{cases} i_1 = I_s(e^{-\alpha V} - 1) \\ i_2 = I_s(e^{\alpha V} - 1) \end{cases} \tag{1}$$

where I_s is the reversed saturation current; α is the slope parameter ($\alpha = \frac{q}{nkT}$), where k is the Boltzmann constant, n is the ideality factor, and T is the operating temperature of the diode.

The conductance of each diode junction is

$$\begin{cases} g_1 = \frac{di_1}{dV} = -\alpha I_s e^{-\alpha V} \\ g_2 = \frac{di_2}{dV} = \alpha I_s e^{\alpha V} \end{cases} \tag{2}$$

The mixing current of the two diode junctions is written as

$$\begin{cases} i_1 = g_1(v_{RF} cos\omega_{RF}t + v_{LO} cos\omega_{LO}t) \\ i_2 = g_2(v_{RF} cos\omega_{RF}t + v_{LO} cos\omega_{LO}t) \end{cases} \tag{3}$$

The total mixing current of the anti-parallel diode pair is

$$\begin{aligned} I_1 = i_1 - i_2 = 2\alpha I_s \sinh(v_{LO} cos\omega_{LO}t) * v_{RF} cos\omega_{RF}t + 2\alpha I_s \cosh(v_{LO} cos\omega_{LO}t) * v_{LO} cos\omega_{LO}t \\ = A\cos\omega_{RF}t + B\cos\omega_{LO}t + C\cos 3\omega_{LO}t + D\cos 5\omega_{LO}t \\ + E\cos(2\omega_{LO}t + \omega_{RF}t) + F\cos(2\omega_{LO}t - \omega_{RF}t) \\ + G\cos(4\omega_{LO}t + \omega_{RF}t) + H\cos(4\omega_{LO}t - \omega_{RF}t) \\ + \cdots + X\cos(m\omega_{LO}t + n\omega_{RF}t) + \cdots \end{aligned} \tag{4}$$

As listed in Table 1, the total mixing current only contains frequency terms $f = mf_{RF} \pm nf_{LO}$, where $m + n$ is odd [19]. The anti-parallel configuration can suppress half of the mixed signal called balanced structure.

The topology of the mixer designed in this paper is presented in Figure 1c. Two Schottky diodes are in anti-series across the LO waveguide [20]. Thus, the two diodes are turned on and off alternately along with the LO signal in TE10 mode. The RF signal is applied to diodes in quasi-TEM mode, where the RF probe is used to transfer TE10 mode to quasi-TEM mode. The mixing current I_2 along the microstrip line is $I_2 = i_4 - i_3$ and outputs only when the phase difference of i_3 and i_4 is π. The output current I_2 is expressed as $I_2 = i_4 + i_3$, and output when i_3 and i_4 are in the same phase.

The currents of junctions can be written as

$$\begin{cases} i_3 = -I_s(e^{-\alpha V} - 1) \\ i_4 = I_s(e^{\alpha V} - 1) \end{cases} \tag{5}$$

The time-varying conduction is

$$\begin{cases} g_3 = \frac{di_3}{dV} = \alpha I_s e^{-\alpha V} \\ g_4 = \frac{di_4}{dV} = \alpha I_s e^{\alpha V} \end{cases} \tag{6}$$

The mixing current of diode junctions is

$$\begin{cases} i_3 = g_3(v_{LO}cos\omega_{LO}t - v_{RF}cos\omega_{RF}t) \\ i_4 = g_4(v_{LO}cos\omega_{LO}t + v_{RF}cos\omega_{RF}t) \end{cases} \tag{7}$$

The mixing current along the microstrip line is

$$\begin{aligned} I_2 = i_4 - i_3 = 2\alpha I_s \cosh(v_{LO}cos\omega_{LO}t) * v_{RF}cos\omega_{RF}t \\ +2\alpha I_s \sinh(v_{LO}cos\omega_{LO}t) * v_{LO}cos\omega_{LO}t \end{aligned} \tag{8}$$

The mixing current of along the LO waveguide is

$$\begin{aligned} I_4 = i_4 + i_3 = 2\alpha I_s \sinh(v_{LO}cos\omega_{LO}t) * v_{RF}cos\omega_{RF}t \\ +2\alpha I_s \cosh(v_{LO}cos\omega_{LO}t) * v_{LO}cos\omega_{LO}t \end{aligned} \tag{9}$$

As listed in Table 1, half of the frequency components can be prevented from output to the microstrip line. However, part of the mixing signal leaks from the LO waveguide and cannot be reused. The leakage results in about 3 dB increment of the conversion loss compared with the anti-parallel mixers in principle. So, the topology is not a balanced structure in Figure 1c.

The advantage of the topology in Figure 1c is that biased mixing can be achieved without using on-chip capacitors, which is a big advantage to sub-harmonic mixers in hybrid integrated in terahertz.

3. Mixer Design

3.1. Mixer Architecture

Figure 2 shows the overall passive circuit structure built in a high frequency structure simulator (HFSS) of the 0.67 THz mixer designed. The LO signal is fed by the WR2.8 rectangular waveguide (711 μm × 356 μm) and reduced the height to 150 μm, while the RF is WR1.5 (381 μm × 191 μm) and reduce the height to 120 μm. Two planar channel Schottky diodes are placed in anti-series across the LO waveguide.

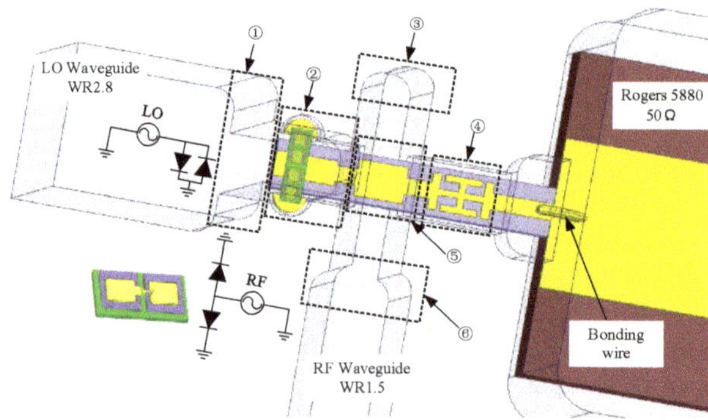

Figure 2. The overall passive circuit modeled in the high frequency structure simulator (HFSS). ① Local Oscillator (LO) reduction waveguide. ② Schottky diode pair with grounded ears. ③ RF waveguide backs short. ④ IF LPF. ⑤ RF probe. ⑥ RF reduction waveguide.

The RF probe is used to coupling the RF signal to the microstrip planar circuit which transfers the TE10 mode to quasi-TEM mode. Due to the orthogonality of TE10 mode and TEM mode, the LO port and RF port are highly isolated. Only one IF LPF (low pass filter) is needed to extract the IF signal and reflect additional harmonics to the diode pair. Due to the difference of mode, the diode pair presents parallel to the LO signal and anti-series to the RF signal. The Rogers 5880 substrate is used as the transition between SMA and quartz substrate.

The mixer has been carefully considered in the following aspects:

- To minimize the size of the mixer, an external Bias-T is used instead of designing inside.
- The cathode pads of the two diodes are directly placed on the pre-designed grounded ears as shown in Figure 3 which can short the LO and RF signal. At the same time, the loop is provided for the IF signal and bias voltage.
- To improve the stability of fabrication and assembly, the Rogers 5880 substrate is used as the transition board to reduce the length of the quartz-glass.
- To enhance the stability of the diode assembly, the shield microstrip line is chosen instead of the suspended microstrip line. The size of the circuit channel is 150 μm × 100 μm which can guarantee the single-mode transmission of the RF signal.

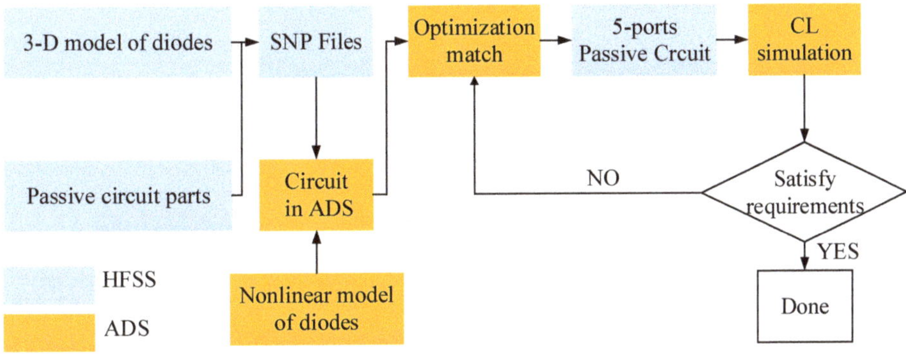

Figure 3. Simulation flow of the 0.67 THz biased mixer.

3.2. Mixer Simulation

Two single-anode Schottky diodes are utilized which are the SD1G2 series produced by Teratech Components Ltd. The diode is based on planar channel structure and the cut-off frequency is about 14 THz which can fully meet the requirement of the 0.67 THz mixer design. Table 2 lists the relative parameters of the Schottky diode.

Table 2. Parameters of the Schottky diode.

Cj0	Rs	Cut-Off Frequency	Overall Size (L × W × H)
0.95 fF	11.5 Ω	14 THz	90 μm × 50 μm × 12 μm

The simulation flow of the 0.67 THz sub-harmonic mixer is presented in Figure 3, which combines the HFSS and Advanced Design Software (ADS). Firstly, in the structure shown in Figure 2, the passive circuit of the mixer is divided into 6 parts. The corresponding scatter-parameters are calculated and exported to SNP files. Second, the complete mixer circuit is built in the ADS, and the harmonic balance algorithm is used to calculate the conversion loss and optimize the matching circuit. Third, the 5-ports overall passive circuit is modeled and simulated in the HFSS and combines the nonlinear diode model in the ADS to verify the final performance. The design is an iterative process.

The circuit in ADS is presented in Figure 4, which is based on the overall optimization method. The SNP 1 to SNP 6 is the corresponding S-parameters of parts in Figure 2. The biased T junction is used to separate the DC voltage and the IF signal and has no insertion loss, which is formed by one ideal capacitor and inductor. The mixing principle is based on the nonlinearity of the Schottky junction which is controlled by the LO signal. According to [21], the LO power is about 6 mw to unbiased sub-harmonic mixers at 670 GHz.

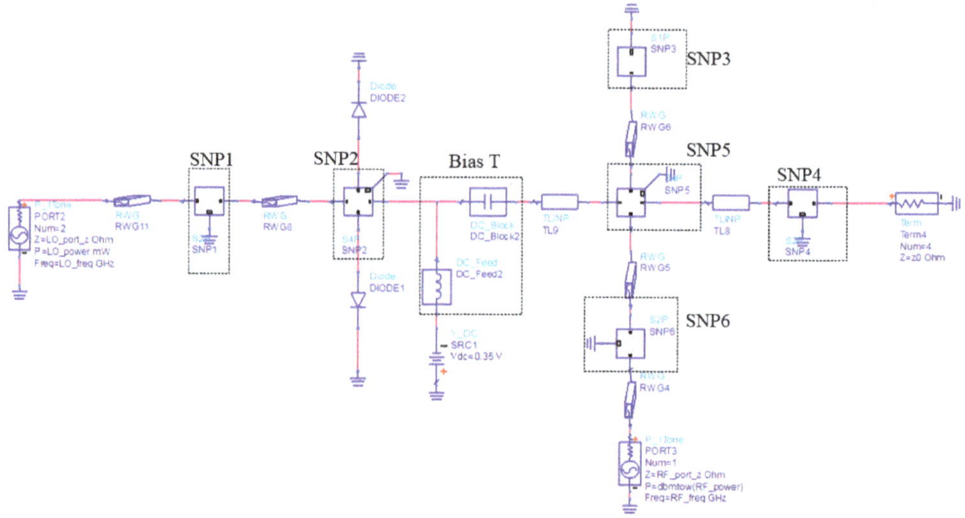

Figure 4. Overall circuit in Advanced Design Software (ADS).

Figure 5 shows the curve of conversion loss under different LO power and DC voltage when the LO and RF frequencies are set to 335 GHz and 671 GHz, respectively. When the LO power is fixed at 2 mw and bias voltage ranges from 0.1 V to 0.9 V, the minimum conversion loss is achieved around the voltage of 0.35 V. Since the driving power is insufficient in the range of 0 to 0.35 V, the conversion loss decreases with the bias voltage increases. However, conversion loss increases in the interval of 0.35 V to 1 V because of the nonlinearity reduction of diodes with the increasing of the bias voltage. When the LO power changes, the phenomenon is similar to the above. If a combination of lower LO power (<2 mw) and higher bias voltage (>0.35 V) is used, the optimal conversion loss cannot achieve because of the dynamic range of the diode junction caused by the LO signal is low.

Figure 6 shows the simulation results when the LO power is 2 mw and the bias voltage is 0.35 mV. The conversion loss is from 10 dB to 12 dB when the RF frequency is range from 653 GHz to 710 GHz. When the LO frequency is varied from 330 GHz to 340 GHz, the change in conversion loss is less than 1 dB. This indicates that the 0.67 THz biased mixer has good RF and LO bandwidth characteristics.

Figure 7 shows the simulated isolation between the RF, LO, and IF ports. Due to the orthogonality of TE10 mode and TEM mode, the isolation of LO port to the RF port is above −50 dB between 300 GHz to 400 GHz, and the isolation of RF to LO is above −29 dB between 650 GHz to 720 GHz. Due to the IF filter, the RF power cannot leak to the IF port, and its isolation is above −18 dB from 650 GHz to 710 GHz.

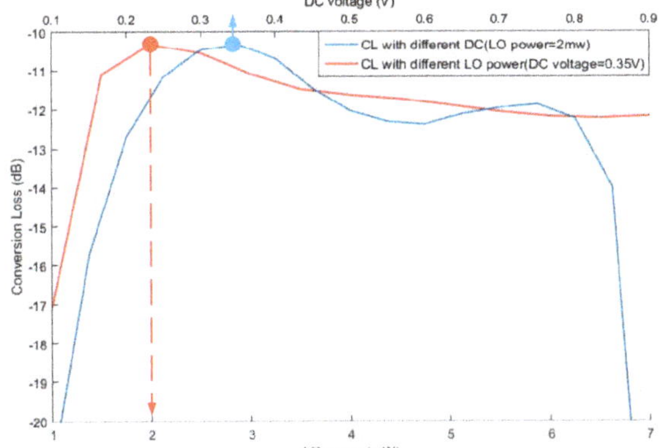

Figure 5. Conversion loss with different LO power and DC voltage.

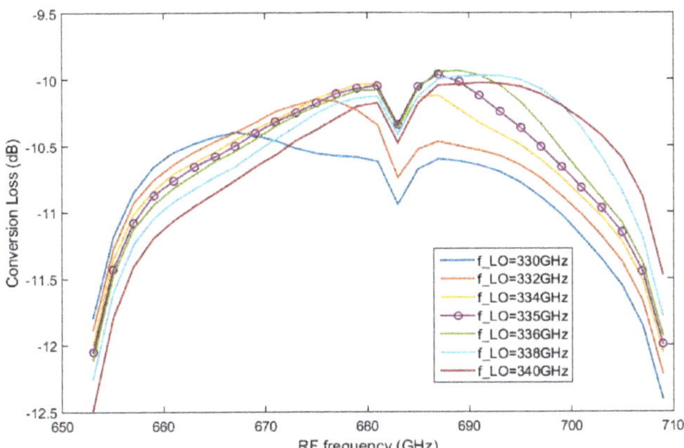

Figure 6. Conversion loss with different LO frequencies.

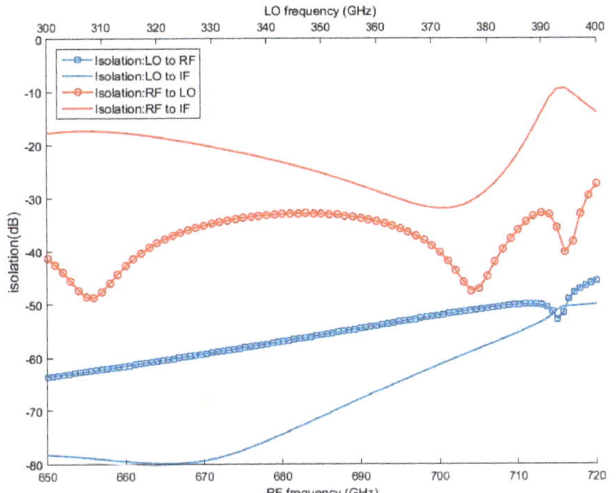

Figure 7. Simulated isolation of RF, LO, and IF ports.

3.3. Mixer Fabrication and Measurement

The cavity of the designed 0.67 GHz biased mixer is made of brass material, and the entire surface is gold plated. The mixer integrates two UG387 flanges for LO and RF waveguides and a female SMA connector for the IF signal. To facilitate assembling, the cavity is cut from the center of the E plane of the waveguide into two parts and locked by screws. As shown in Figure 8, the quartz-glass substrate and diodes are fixed by conductive adhesive, and the Rogers 5880 substrate is glued by tin solder.

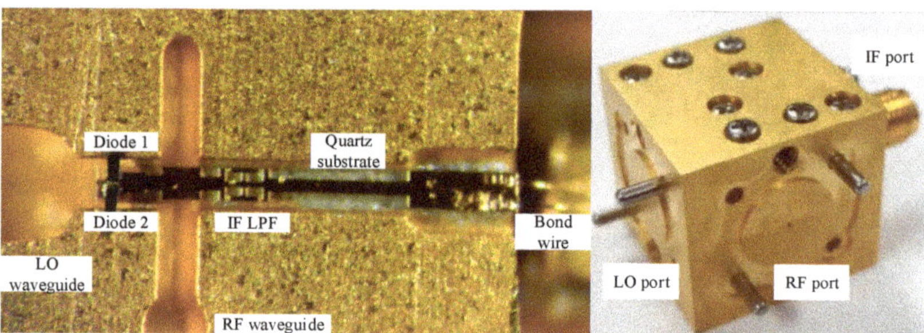

Figure 8. Circuit in the lower block of the mixer; assembled mixer.

Figure 9 shows the test diagram of the mixer. The RF and LO signals are generated by two different links, and the power and spectrum of the IF signal are measured by a spectrum analyzer. An external Bias T is used to separate the DC and IF signals. The LO signal is generated by a multiplier chain which is composed of a signal generator, W band multiplier, W band PA, and 330 GHz doubler. The RF signal is and generated by a signal generator and * 54 multiplier module produced by VDI.

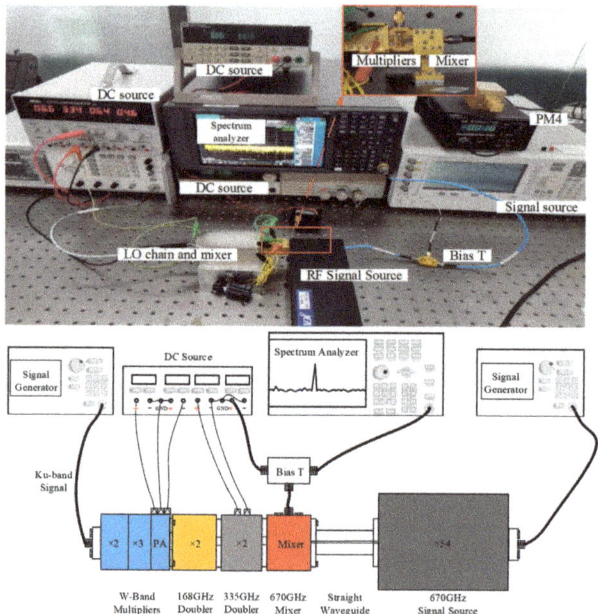

Figure 9. Test bench and measurement diagram of the mixer.

Before measuring the conversion loss, the respective output power of the LO and RF chains needs to be measured and calibrated by a PM4 power meter. The maximum LO power generated by the LO chain is about 13 mW between 330 GHz and 340 GHz. The RF signal power is between −18 dBm to −15 dBm from 600 GHz to 700 GHz.

Figure 10 presents the Signal Side-Band (SSB) conversion loss versus RF frequency when the LO is fixed at 2 mw@335 GHz. The conversion loss shown has been corrected for attenuation of cables and the insertion loss of the Biased T. The best measured SSB conversion loss is 15.3 dB@667 GHz. In the RF band of 650 GHz–690 GHz, the conversion loss is below 20 dB, besides the frequency of 679 GHz and 685 GHz. The typical SSB conversion loss is 18.2 dB.

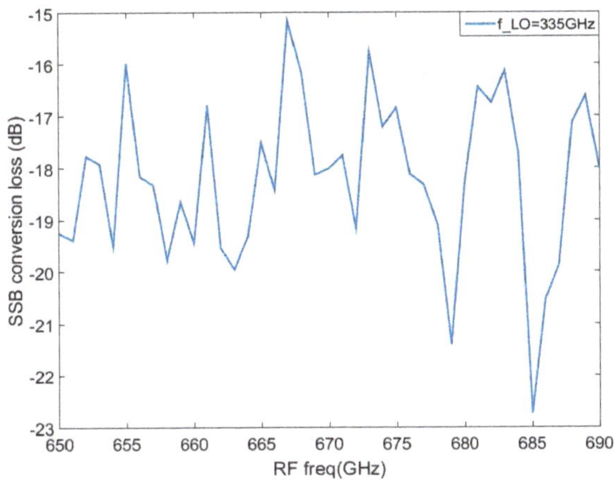

Figure 10. Measured Signal Side-Band (SSB) conversion loss.

4. Discussion

As shown in Figure 10, the curve of the SSB conversion loss has a large jitter in the band of 670 GHz to 690 GHz than in 650 GHz to 670 GHz. The phenomenon can be explained as follows. First, the RF power is measured by the PM4 power meter which is a thermal power meter that calculates the total power of input signals. While the output RF power ranges from −20 dBm (10 mW) to −18 dBm (15.8 mW) in 670 GHz to 690 GHz, which cannot be measured accurately. The 3 mW to 5 mW measurement error of the PM4 power meter is normal which may lead conversion loss error of 2 dB. Second, the RF source utilized is a frequency multiplier module (* 54) which has many harmonic components that affect the RF power.

As can be seen from Figures 6 and 10, there is a conversion loss gap between simulation and measurement. The gap is caused by the following reasons. First, the nonlinear model of the Schottky diode junction utilized in simulation is provided by the ADS software, which is a P-N junction model but not Schottky diode junction (metal-semiconductor junction). When the working frequency reaches 0.6 THz and above, the current saturation in the diode junction and planar structure can increase the conversion loss [21]. It is urgent to modify the diode junction model in the simulation. Second, the assembling error of Schottky diodes and the substrate deteriorates the performance, which is caused by the thickness of conductive adhesive and the alignment of the diodes. In the assembling, the alignment error between the two diodes is about 20 μm, which leads to unbalanced mixing and deteriorate the conversion loss. Besides, the conductivity of conductive adhesive is not ideal and cannot be accurately characterized in simulation, as the conductive adhesive is the self-made mixture of epoxy and silver.

Table 3 shows several reported sub-harmonic mixers and receivers working around 600 GHz. The mixer reported in Reference [10,22] are all based on anti-parallel Schottky diode pair and hybrid integration. Thus, the LO power is higher than the mixer designed in this paper. The 0.67 THz biased sub-harmonic mixer has better conversion loss than the mixer in [21]. The mixer reported in Reference [13] is based on advanced membrane monolithic integration technology and anti-parallel architecture. Thus, it has better conversion loss than the mixer design.

Table 3. Summary of published terahertz mixer working around 600 GHz.

Frequency (GHz)	Biased	LO Power (mw)	Conversion Loss (dB)	Structure	Reference
638–715	No	2–8	8.2–12 (DSB)	Hybrid	[10]
520–590	Yes		10.6–11.7 (DSB)	Monolithic	[13]
660–710	No	6	13–20 (DSB)	Hybrid	[22]
650–690	Yes	3	Optimum: 15.3 (SSB) Typical: 18.2 dB	Hybrid	This work

Compared with those mixers, the 0.67 THz biased sub-harmonic mixer has disadvantages and advantages as follows. In terms of conversion loss, the anti-series topology utilized has intrinsically defective in suppressing harmonics compared with anti-parallel topology, as analyzed in Section 2. But the anti-series topology can be used to biased mixers in the hybrid integration that effectively reduce the LO power. The performance can be improved in two ways in simulation. First, the Voltage Standing Wave Ratio (VSWR) of the LO port needs to be further improved to reduce the LO power. Second, the anti-series Schottky diode pair can be used to replace the two discrete diodes. The diode pair can eliminate the alignment error of the discrete diodes in assembling.

5. Conclusions

A 0.67 THz biased sub-harmonic mixer in hybrid integration has been designed and measured based on an anti-series Schottky diode placed across the LO waveguide. The circuit topology utilized is detailed, analyzed and compared with traditional anti-parallel configurations. The mixer design can realize the biased mixing structure without using discrete chip capacitors, and high isolation between RF, LO, and IF ports. The measured data shows the optimum SSB conversion loss is 15.3 dB at 667 GHz. In the RF frequency band of 650 GHz to 690 GHz, the typical value of conversion loss is 18.2 dB. The mixer design effectively decreased the LO power in the hybrid integrated Schottky diode-based sub-harmonic mixer. At the same time, the 0.67 THz biased sub-harmonic mixer has great prospects in ice cloud detection and planetary exploration.

Author Contributions: Conceptualization, methodology, software, and writing, G.J., D.Z., J.M., and C.Y.; investigation, S.L.; visualization, G.J. All authors have read and agreed to the published version of the manuscript.

Funding: This research received no external funding.

Conflicts of Interest: The authors declare no conflict of interest.

References

1. Siegel, P.H. Terahertz technology. *IEEE Trans. Microw. Theory Tech.* **2002**, *50*, 910–928. [CrossRef]
2. Meng, J.; Zhang, D.H.; Ji, G.Y.; Yao, C.F.; Jiang, C.H.; Liu, S.Y. Design of a 335 GHz Frequency Multiplier Source Based on Two Schemes. *Electronics* **2019**, *8*, 948. [CrossRef]
3. Mehdi, I.; Siles, J.V.; Maestrini, A.; Lin, R.; Lee, C.; Schlecht, E. Chattopadhyay. Local oscillator sub-systems for array receivers in the 1–3 THz range. In Proceedings of the SPIE Spie Astronomical Telescopes + Instrumentation, Amsterdam, The Netherlands, 24 September 2012.
4. Maestrini, A.; Thomas, B.; Wang, H.; Jung-Kubial, C.; Treuttel, J.; Jin, Y.; Chattopadhyay, G.; Mehdi, I.; Beaudin, G. Schottky diode-based terahertz frequency multipliers and mixers. *C. R. Phys.* **2010**, *11*, 480–495. [CrossRef]

5. Waliwander, T.; Crowley, M.; Fehilly, M.; Lederer, D.; Pike, J.; Floyd, L.; O'Connel, D. Sub-millimeter Wave 183 GHz and 366 GHz MMIC membrane sub-harmonic mixers. In Proceedings of the 2011 IEEE MTT-S International Microwave Symposium, Baltimore, MD, USA, 5–10 June 2011.
6. Chen, Z.; Zhang, B.; Fan, Y.; Yuan, Y. Design of a low noise 190–240 GHz subharmonic mixer based on 3D geometric modeling of Schottky diodes and CAD load-pull techniques. *IEICE Electron. Express* **2016**, *13*, 20160604. [CrossRef]
7. Thomas, B.; Maestrini, A.; Beaudin, G. A Low-Noise Fixed-Tuned 300–360 GHz Sub-Harmonic Mixer Using Planar Schottky Diodes. *IEEE Microw. Wirel. Compon. Lett.* **2005**, *15*, 865–867. [CrossRef]
8. Tong, X.D.; Li, Q.; An, N.; Wang, W.J.; Deng, X.D.; Zhang, L.; Liu, H.T.; Zeng, J.P.; Li, Z.Q.; Tang, H.L.; et al. The Study of 0.34GTHz Monolithically Integrated Fourth Subharmonic Mixer Using Planar Schottky Barrier Diode. *J. Infrared Millim. Terahertz Waves* **2015**, *36*, 1112–1122. [CrossRef]
9. Yang, F.; Meng, H.F.; Duo, W.B.; Sun, Z.L. Terahertz Sub-harmonic Mixer Using Discrete Schottky Diode for Planetary Science and Remote Sensing. *J. Infrared Millim. Terahertz Waves* **2017**, *38*, 630–637. [CrossRef]
10. He, Y.; Tian, Y.; Miao, L.; Jiang, J.; Deng, X.J. A Broadband 630–720 GHz Schottky Based Sub-Harmonic Mixer Using Intrinsic Resonances of Hammer-Head Filter. *China Commun.* **2019**, *16*, 86–94.
11. Maestrini, A.; Ward, J.S.; Gill, J.J.; Javadi, H.S.; Schlecht, E.; Tripon-Canseliet, C.; Chattopadhyay, G.; Mehdi, I. A 540–640-GHz high-efficiency four-anode frequency tripler. *IEEE Trans. Microw. Theory Tech.* **2005**, *53*, 2835–2843. [CrossRef]
12. Maestrini, A.; Ward, J.S.; Gill, J.J.; Lee, C.; Thomas, B.; Lin, R.H.; Chattopadhyay, G.; Mehdi, I. A Frequency-Multiplied Source with More Than 1 mW of Power Across the 840–900-GHz Band. *IEEE Trans. Microw. Theory Tech.* **2010**, *58*, 1925–1932. [CrossRef]
13. Schlecht, E.; Gill, J.; Dengler, R.; Lin, R.; Tsang, R.; Mehdi, I. A Unique 520-590 GHz Biased Subharmonically-Pumped Schottky Mixer. *IEEE Microw. Wirel. Compon. Lett.* **2007**, *17*, 879–881. [CrossRef]
14. Thomas, B.; Maestrini, A.; Matheson, D.; Mehdi, I.; de Maagt, P. Design of an 874 GHz biasable sub-harmonic mixer based on MMIC membrane planar schottky diodes. In Proceedings of the 2008 33rd International Conference on Infrared, Millimeter and Terahertz Waves, Pasadena, CA, USA, 15–19 September 2008.
15. Hammar, A.; Sobis, P.; Drakinskiy, V.; Emrich, A.; Wadefalk, N.; Schleeh, J.; Stake, J. Low noise 874 GHz receivers for the International Submillimetre Airborne Radiometer (ISMAR). *Rev. Sci. Instrum.* **2018**, *89*, 055104. [CrossRef] [PubMed]
16. Schlecht, E.; Siles J, V.; Lee, C.; Lin, R.; Thomas, B.; Chattopadhyay, G.; Medhi, I. Schottky Diode Based 1.2 THz Receivers Operating at Room-Temperature and Below for Planetary Atmospheric Sounding. *IEEE Trans. Terahertz Sci. Technol.* **2014**, *4*, 661–669. [CrossRef]
17. Thomas, B.; Siles, J.; Schlecht, E.; Maestrini, A.; Chattopadhyay, G.; Lee, C.; Jung-Kubiak, C.; Mehdi, I.; Gulkis, S. First results of a 1.2 THz MMIC sub-harmonic mixer based GaAs Schottky diodes for planetary atmospheric remote sensing. *ISSTT Proc.* **2012**, 100–102. Available online: http://https://www.nrao.edu/meetings/isstt/2012.shtml (accessed on 10 January 2020).
18. Treuttel, J.; Schlecht, E.; Siles, J.; Lee, C.; Lin, R.; Thomas, B. A 2 THz Schottky solid-state heterodyne receiver for atmospheric studies. In Proceedings of the SPIE Millimeter, Submillimeter, and Far-Infrared Detectors and Instrumentation for Astronomy VIII, Edinburgh, UK, 16 June 2016.
19. Cohn, M.; Degenford, J.E.; Newman, B.A. Harmonic Mixing with an Anti-Parallel Diode Pair. In Proceedings of the 1974 IEEE S-MTT International Microwave Symposium, Atlanta, GA, USA, 12–14 June 1975.
20. Schlecht, E.; Gill, J.; Siegel, P.H.; Oswald, J.; Mehdi, I. Novel Designs for Submillimeter Subharmonic and Fundamental Schottky Mixers. In Proceedings of the 14th International Symposium on Space Terahertz Technology, Arizona, AZ, USA, 22–24 April 2003.
21. Crowe, T.W.; Mattauch, R.J.; Roser, H.P.; Bishop, W.L.; Peatman, W.C.B.; Liu, X.L. GaAs Schottky diodes for THz mixing applications. *Proc. IEEE* **1992**, *80*, 1827–1841. [CrossRef]
22. Jiang, J.; He, Y.; Wang, C.; Liu, J.; Tian, Y.L.; Zhang, J.; Deng, X.J. 0.67 THz sub-harmonic mixer based on Schottky diode and hammer-head filter. *J. Infrared Millim. Terahertz Waves* **2016**, *35*, 418–424.

Article

Correction of Optical Delay Line Errors in Terahertz Time-Domain Spectroscopy

Alexander Mamrashev [1,*], Fedor Minakov [1,2], Lev Maximov [1,2], Nazar Nikolaev [1] and Pavel Chapovsky [2,3]

1 Laboratory of Information Optics, Institute of Automation and Electrometry SB RAS, 630090 Novosibirsk, Russia; minakovfa95@gmail.com (F.M.); lev.maximov@gmail.com (L.M.); nazar@iae.nsk.su (N.N.)
2 Physics Department, Novosibirsk State University, 630090 Novosibirsk, Russia
3 Laboratory of Nonlinear Spectroscopy of Gases, Institute of Automation and Electrometry SB RAS, 630090 Novosibirsk, Russia; chapovsky@iae.nsk.su
* Correspondence: mamrashev@iae.nsk.su; Tel.: +7-383-330-8378

Received: 24 October 2019; Accepted: 22 November 2019; Published: 26 November 2019

Abstract: One of the key elements of terahertz time-domain spectrometers is the optical delay line. Usually it consists of a motorized translation stage and a corner reflector mounted on its top. Errors in the positioning of the translation stage lead to various distortions of the measured waveform of terahertz pulses and, therefore, terahertz spectra. In this paper, the accuracy of position measurements is improved by using an optical encoder. Three types of systematic errors are found: Increasing and periodic offsets of the translation stage position, as well as a drift of its initial position in a series of consecutive measurements. The influence of the detected errors on the measured terahertz spectra is studied and correction methods are proposed.

Keywords: terahertz spectroscopy; optical delay line; correction; optical encoder; terahertz spectra; terahertz metrology

---

1. Introduction

Terahertz time-domain spectroscopy (THz-TDS) has become one of the most common methods for studying optical and dielectric properties of materials in the frequency range of 0.1–10 THz with the development of femtosecond laser technology [1,2]. This method is used to study nonlinear optical crystals [3,4], nuclear spin isomers [5], complex biomolecules [6], and charge carriers in solids [7,8].

The principle of operation of terahertz time-domain spectrometers is based on the generation of terahertz pulses and measurement of their electric field waveform. An optical delay line allows changing of the path difference between generation and detection optical channels of the spectrometer. This enables point-by-point sampling of the terahertz pulse waveform. Digital Fourier transform is used to calculate the spectra of the measured signal that contain information on the absorption coefficient and the refractive index of the media under study. Details of the spectrometer operation will be discussed later.

Various elements of spectrometers exhibit random and systematic errors, leading to distortion of terahertz pulses and, therefore, terahertz spectra [9–13]. The errors in the amplitude of terahertz pulses are mainly determined by random fluctuations and the long-term drift of the THz generation system, which consists of a femtosecond pump laser and an optical-to-terahertz converter. Errors in the positioning of the optical delay line lead to more complex distortions. In [14], the effect of a drift of the initial position of the delay line in a series of sequentially measured terahertz pulses was considered. It was shown that it led to error proportional to the THz signal shifted by a quarter cycle. Spectroscopy of thin films and attenuated total internal reflection spectroscopy are especially sensitive to such error [15]. In [16], random errors of optical delay line positioning were considered. It was shown by Monte Carlo modeling that the terahertz spectrum was impaired by spectrally independent additive noise, which was directly proportional to the noise in the time domain and the square root of the sampling time steps of THz pulses. It was shown in [17] that periodic sampling error results in the presence of spurious mirror copies of the main pulse spectra, in which frequency and amplitude depend on the period and amplitude of the error, respectively.

Optical delay lines in THz spectrometers are usually based on corner reflectors on mechanical translation stages driven by voice coils or stepper motors. There are also conceptually different non-mechanical approaches for shifting time delay between generation and detection optical channels such as ECOPS and ASOPS techniques [18,19]. Imperfections in mechanical translation stages lead to positioning errors that are usually not considered by THz-TDS operators. However, these errors can be eliminated by software or hardware solutions. Algorithms are used to correct the time shift in a series of measured THz pulses [20,21]. Interferometers are used to more accurately measure the delay line position [22]. However, they require additional laser, optical elements and electronics making overall setup more expensive and difficult to operate.

In this paper, we propose a simple and cost-effective method to improve the accuracy of optical delay line positioning by using an optical encoder. We detect systematic positioning errors, study their effects on the measured terahertz spectra, and propose correction methods. The studies are conducted on a custom-made THz spectrometer with a delay line based on a motorized translation stage upgraded with the optical encoder.

2. Experimental Setup and Measurement Procedure

2.1. THz Spectrometer

To study the random and systematic errors of the optical delay line and the effect of these errors on the measured terahertz spectra, experiments were carried out on a THz spectrometer created in the Institute of Automation and Electrometry of Siberian Branch of Russian Academy of Sciences. The experimental setup is a standard terahertz time-domain spectrometer (Figure 1). The source of the femtosecond pump and probe pulses is an Er-doped fiber laser with a second harmonic generation module FFS-SHG (Toptica Photonics AG, Munich, Germany) providing radiation with the following parameters: Central wavelength—775 nm, pulse duration—130 fs, and mean power—80 mW. A photoconductive antenna iPCA-21-05-300-800-h with microlense array and interdigitated electrodes on semi-insulating GaAs substrate (Batop GmbH, Jena, Germany) serve as a THz generator. The terahertz field is detected by electro-optic sampling in a (110)-cut 2 mm ZnTe crystal. In this method, the probe pulse passes through ZnTe crystal that became birefringent under terahertz electric field and changes polarization. A quarter-wave plate and a Wollaston prism split the probe pulse into two beams with orthogonal linear polarizations. A pair of photodiodes detect the difference between the powers of two beams which is proportional to the electric field of the terahertz radiation. The differential signal is measured by a lock-in amplifier SR830 (Stanford Research Systems, Sunnyvale, CA, USA) tuned to the frequency of ~8 kHz provided by an iPCA voltage generator.

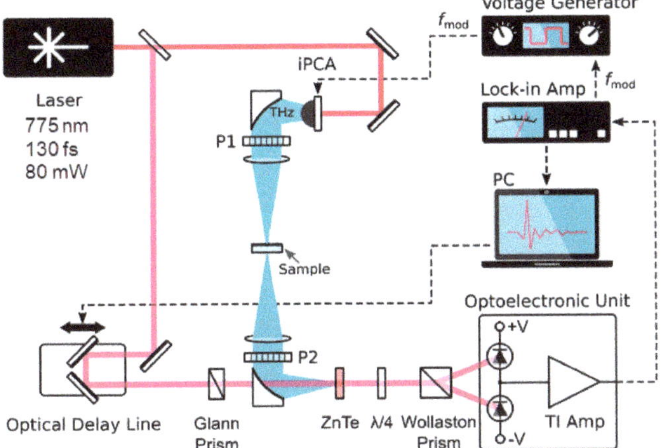

Figure 1. Scheme of the custom-made terahertz time-domain spectrometer.

The optical path lengths of generation and detection arms are equal so that the terahertz pulse and probe laser pulse synchronously arrive at the detection crystal. The optical delay line allows changing of the optical path length of the detection arm and sampling of the terahertz pulse waveform. The Fourier transform of the measured waveform gives its spectrum. The delay line is a motorized translation stage 8MT173-50-20 (Standa Ltd., Vilnius, Lithuania) with a corner reflector mounted on its top. For such setup movement of the translation stage by a distance of Δl corresponds to a time delay of $\Delta t = 2 \cdot \Delta l / c$, where c is the speed of light.

2.2. Upgraded Optical Delay Line

The leadscrew of the translation stage 8MT173-50-20 is driven by a stepper motor operated by an 8SMC1-USBhF motion controller. Full step resolution of the motor is $\Delta l = 1.25$ μm which corresponds to a time delay of $\Delta t = 8.34$ fs. The controller provides resolution down to 1/8 of the step, speed up to 5 mm/s, long-range movements with programmable acceleration and deceleration. One revolution of the screw corresponds to a movement of $\Delta l_1 = 250$ μm, i.e., $\Delta t_1 = 1.67$ ps in the time domain. The full movement range is $L = 50$ mm, i.e., $T = 334$ ps. The controller of the stage measures only its relative position (there are no absolute data) by counting the number of steps done by a stepper motor.

To study random and systematic errors of the optical delay line, a Resolute RL32BAT001B50 optical encoder with an absolute scale RTLA (Renishaw, Gloucestershire, United Kingdom) was installed on the translation stage (Figure 2). The readhead of the optical encoder scans the scale with a code allowing determination of the absolute position of the readhead relative to the scale. The scale is a low-profile stainless-steel tape. The accuracy of the scale is ± 5 μm/m, the coefficient of thermal expansion at the temperature of 20 °C is 10.1 ± 0.2 μm/m/°C. The resolution and sub-divisional error of the position measurements are 1 nm and ± 40 nm, respectively. For our task, a scale segment of 10 cm is cut, hence its absolute positioning accuracy is 0.5 μm, and the coefficient of thermal expansion is 1.01 μm/°C. These values correspond to the time domain absolute accuracy of ± 3.34 fs and random error of ± 0.27 fs. The readhead and its communication protocol support the sampling rate up to 25 kHz.

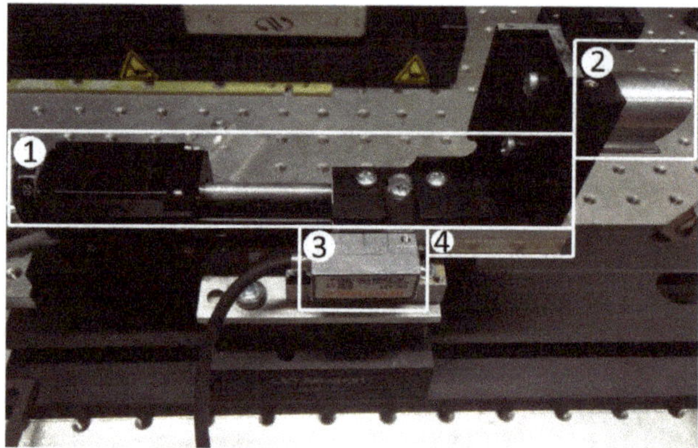

Figure 2. Upgraded optical delay line consisting of a translation stage (1), a corner reflector (2), an optical encoder readhead (3), and a scale (4).

The design of the mounting of the optical encoder and the scale allows us quick exchange between four analogous translation stages for testing of their accuracies. The design also allows us to adjust the readhead inclination and its position in two axes and set the encoder to the optimal reading state. After adjustment, the readhead is fixed, and the scale mounted on the translation stage platform moves relative to the readhead.

2.3. Measurement Procedure

The measurements of the translation stage position were performed in a point-by-point manner. The translation stage speed was set to a default value of 0.78 mm/s. The measurements were performed with the step of 12.5 μm in the range of 45 mm. The step size of 12.5 um contained integer number of stepper motor full steps. When moving the translation stage, we recorded two sets of data: The position of the translation stage measured by counting stepper motor steps (L_{st}) and the position obtained from the optical encoder (L_{en}). Values measured at the initial position of the translation stage were used as a reference (zero) position. A comparison of the L_{st} and L_{en} positions was carried out for four analogous translation stages 8MT173-50-20 that are used in our terahertz spectrometer A comparison of the positions was carried out when the translation stages moved both in the positive and in the negative directions. No significant differences were found in these cases; therefore, further results are presented only for positive movement direction.

To assess the repeatability of the results an additional series of measurements for one of the translation stages positions was carried out. In this series, after each measurement, the translation stage returned to the zero starting position according to the value of L_{st}. The translation stage position was measured together with the terahertz pulse waveform. The lock-in amplifier time constant was set to 100 ms. We waited the amount of time required for the signal settlement at each point before saving the terahertz field measurement.

3. Results

Figure 3 shows the difference between the optical encoder measurements L_{en} and the positions of the four translation stages L_{st}. All of them exhibit similar behavior. Two types of systematic errors can be seen in the figure: An increasing position difference and a periodic error. There are also sharp jumps in position due to scratches, damages, and other mechanical defects of the translation stage screw. Figure 3a shows an increasing offset of position relative to the more accurate data from the optical encoder for all four translation stages. Over the entire range, this offset has a magnitude of the order of tens of micrometers that significantly exceeds the encoder accuracy of 0.5 μm. As we zoom in (see Figure 3b) we can see some periodic position mismatch. The amplitude of the oscillations varies from 0.4 to 1 μm for different translation stages. The oscillation period is 250 μm, which corresponds to the full revolution of the translation stage screw.

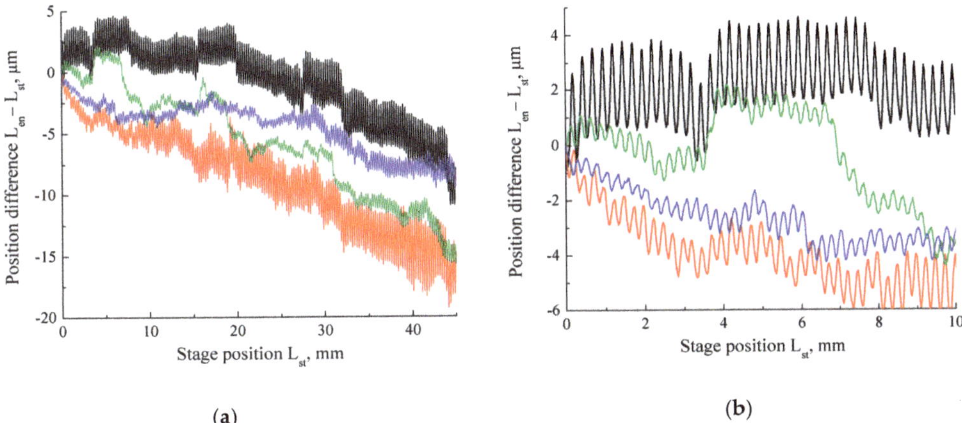

Figure 3. The difference between the translation stage positions and the optical encoder measurements for four translation stages in the ranges: (**a**) 45 mm; (**b**) 10 mm.

The third type of systematic errors, a drift of the initial position is detected by repeated measurements of the first translation stage. Figure 4 shows the difference between the optical encoder measurements L_{en} and the translation stage position L_{st} in a series of ten consecutive measurements. In the series, after each measurement, the translation stage was supposed to return to the zero starting position. However, Figure 4 shows that the initial position drifted according to the encoder measurements L_{en}. The position shift was about 1 μm (corresponding to the time shift of 6.7 fs) between the first measurements and gradually decreased in the subsequent measurements in the series.

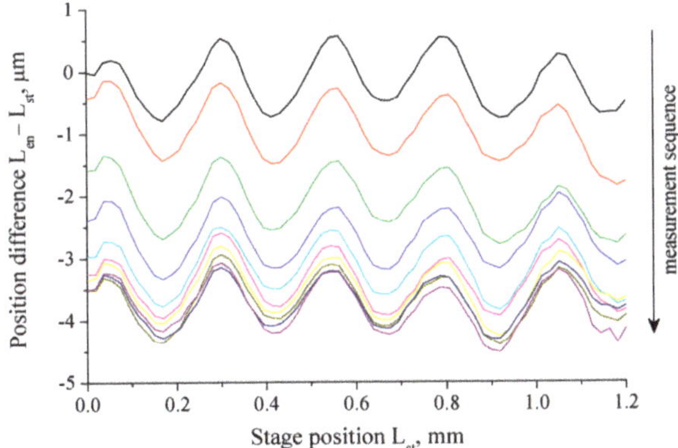

Figure 4. The drift of the initial position measured by the optical encoder in a series of ten consecutive measurements (each is depicted with its own color). Sequence of measurements is from top to bottom.

4. The Effect of Translation Stage Errors on Terahertz Spectra

For a start, an assessment of the distortions caused by the increasing position offset, i.e., a systematic error of the first type is made. We estimate the effect for one of the stages as an example. It has an offset of 16 μm in the range of 40 mm (see the red line in Figure 3a). The observed offset can be approximated by a straight line with a slope of $\alpha = -0.0004$. Let $E(t_{st})$ denote the amplitude of the terahertz signal. It depends on the time delay corresponding to the position of the translation stage. Time delay measured by the optical encoder t_{en} can be expressed with the formula $t_{en} = t_{st} + \alpha t_{st}$. Therefore, the amplitude of the THz pulse equals $E(t_{en}/(1+\alpha)) = E(\alpha_1 t_{en})$, where $\alpha_1 = 1.0004$. We can see that the measured time-domain signal is stretched by the factor of α_1 along the time axis. From the properties of the Fourier transform, it is clear that the spectrum is linearly squeezed along the frequency axis by the same factor. For example, for the frequency of 1 THz the corresponding frequency shift is 400 MHz. Such distortion becomes especially noticeable in spectra with narrow absorption lines.

Let us consider the influence of the second type of systematic error, i.e., oscillating mismatch between the optical encoder measurements and the position of the translation stage (see Figure 3b). According to [17], periodic positioning error leads to the emergence of spurious spectra. These spectra are mirror copies of the true spectrum of the THz pulse emerging around the frequency of the periodic error and its harmonics. In our case, the oscillation period of 250 μm corresponds to 1.67 ps in the time domain, which leads to the appearance of spurious spectra around the frequency of 600 GHz and its harmonics. The amplitude of the spurious spectra is proportional to the amplitude (~5 fs) and the frequency (600 GHz) of the periodic error according to the theoretical estimations in [17]. Thus, the proportionality coefficient can be calculated $\approx 3 \cdot 10^{-3}$. The obtained value is comparable with the signal-to-noise ratio of the THz spectrometer, which makes it difficult to detect.

The third error, the drift of the initial position of the translation stage in a series of measurements, affects averaging over a series of terahertz spectra [14]. In this case, an increase in the number of measurements does not lead to an increase in the signal-to-noise ratio [10].

5. Correction

More accurate data from the optical encoder can be used to correct systematic errors of the delay line as well as errors due to the mechanical defects in the translation stage screw. For this, the THz electric field sampled at each position of the delay line is associated with the position more accurately measured by the optical encoder. Then by interpolation, the equidistance of the sampling positions is restored. As a result, the corrected waveform of the terahertz pulse is based on the positions measured with the optical encoder and does not contain systematic errors associated with the translation stage. The results of the correction algorithm that eliminates the systematic errors of the first and the second types are presented in Figures 5 and 6, respectively.

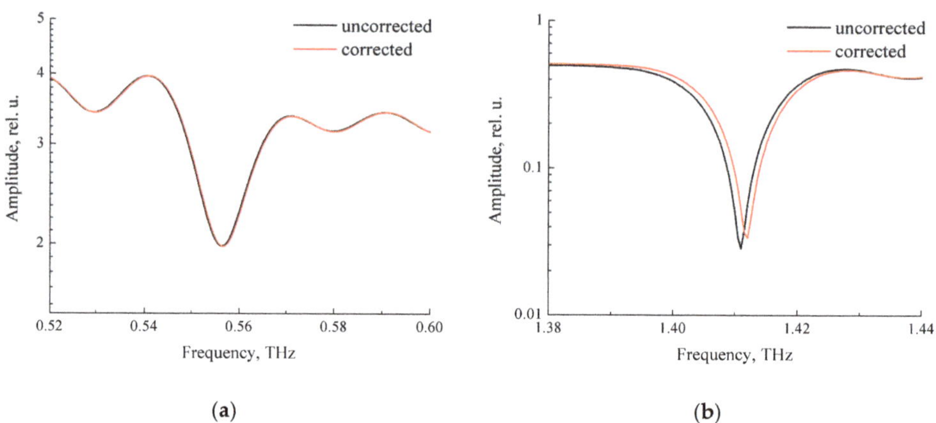

Figure 5. Terahertz (THz) spectra near two water vapor absorption lines before (black line) and after (red line) applying the correction: (**a**) 0.557 THz (**b**) 1.411 THz.

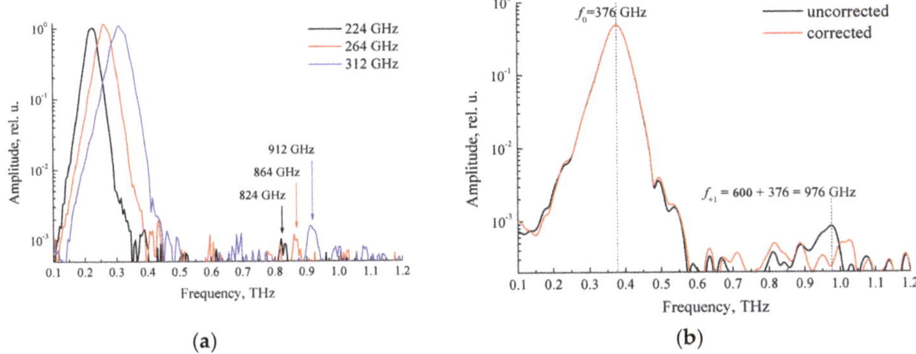

Figure 6. Spectra of THz pulses passed through bandpass filters with the central frequency: (**a**) 224 GHz (black line), 264 GHz (red line), and 312 GHz (blue line), arrows indicate the frequencies of the spurious spectra (**b**) 376 GHz before (black line) and after (red line) correction.

In Figure 5 one can see an example of distortion caused by the error of the first type. We measure terahertz spectrum in the atmosphere containing water vapor that has many rotational absorption lines in the terahertz spectral range [5]. Thus, we can observe the effect of the error by studying THz spectra near two absorption lines of water vapor at 0.557 THz (Figure 5a) and at 1.411 THz (Figure 5b). As expected from the analysis of this error, the spectral shift increases with increasing frequency. It amounts to ~500 MHz for the line at 1.411 THz. Corrected spectra appear to be closer to the tabulated values of water absorption lines in HITRAN database [23] than the uncorrected ones.

We test the influence of the systematic error of the second type by measuring spectra of terahertz pulses passing through high-contrast quasi-optical bandpass filters. The filters are designed as multilayer frequency selective surfaces produced with technologies of photolithography and electroplating [24,25]. They were originally developed for spectro-radiometric applications in electron-beam-plasma experiments on generating high-power sub-terahertz radiation [25,26]. The filters have a bandwidth of 12%–20% and out-of-band transmission of ~10^{-4}. Such filters provide us with an opportunity to clearly observe spurious terahertz spectra since they have a pronounced peak and near-zero out-of-band transmission.

Figure 6a shows the spectra of terahertz pulses passing through high-contrast bandpass filters with the center frequencies of 224, 264, and 312 GHz. It can be seen how the systematic error of the second type leads to the emergence of spurious spectra. The central frequencies of the false spectral features are 600 ± ν where 600 GHz is the frequency of the periodic error and ν is the filter frequency. Figure 6b shows the spectra for 376 GHz filter before and after correction. Its amplitude is ~580 times smaller than the amplitude of the main spectral feature and is practically at the noise level of the spectrometer.

Lastly, we consider the systematic error of the third type. The drift of the initial position by 1 μm measured by the optical encoder (see Figure 4) should correspond to the time shift of 6.7 fs. However, such shift is barely noticeable in the waveforms of consecutively measured terahertz pulses (see Figure 7).

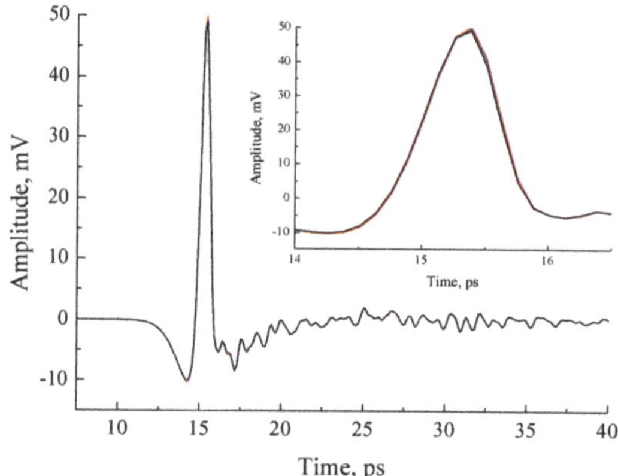

Figure 7. Waveforms of ten consecutively measured THz pulses presented in the ranges of 5–40 ps (main figure) and 14–16.5 ps (inset).

For clarity, we compare the initial position drift measured by encoder and time drift of THz pulses on the same scale in Figure 8. The drifts are calculated relative to the first measurement in the series. It can be seen that it takes more than 5 measurements until encoder position drift settles down at −3.5 µm. In contrast, time drift of THz pulses is significant only between the first and the second measurements due to translation stage backlash. Apparently, this discrepancy is associated with the heating of the stepper motor and translation stage platform during operation and the thermal expansion of the optical encoder scale. Heating by 1 °C leads to a shift of ~1 µm at the free end of the scale. As a result, setting the initial position based on counting steps of the translation stage stepper motor turns out to be more accurate than based on the optical encoder data. Residual THz time drift can be algorithmically compensated by time shift and interpolation of the signals [20,21].

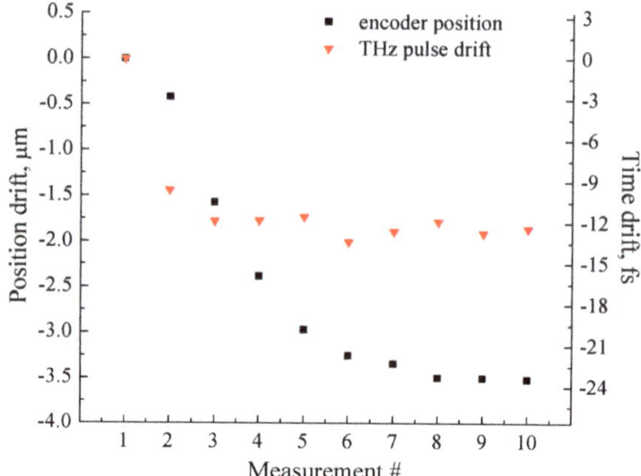

Figure 8. The drift of the initial position measured by the optical encoder (black squares) and measured based on the time shift of terahertz pulses (red triangles) relative to the first measurement.

6. Conclusions

In the work, we upgraded translation-stage-based optical delay line of the terahertz time-domain spectrometer. Additionally, we installed the optical encoder that made it possible to get more accurate information on the position of the translation stage during the terahertz measurements. We were able to study systematic errors associated with the translation stage without the optical encoder. We also used encoder positioning data to programmatically correct these errors. The proposed method is similar to interferometry-aided terahertz spectroscopy. However, the encoder is cheaper, easier to install and operate while having comparable accuracy.

Using the encoder, we detected three different types of systematic errors in the optical delay line. Firstly, an increasing difference in positions measured by the stepper motor of the translation stage and the optical encoder. This error led to a linear distortion of terahertz spectra along the frequency axis, which is especially critical in gas spectroscopy. Secondly, a periodic positioning error associated with the revolution of the translation stage screw was observed. It was shown that this error led to the appearance of spurious spectra having amplitude on the order of ~10^{-3} compared to the main pulse spectrum near the frequency of 600 GHz. The appearance of the spurious spectrum was demonstrated by studying the transmission of a band-pass filter with a central frequency of 376 GHz. Thirdly, in a series of ten consecutive measurements of terahertz pulses with the same delay line, a shift in the initial position of the translation stage was detected. Apparently, this effect appeared due to the heating of the stepper motor and translation stage platform during operation, which led to the heating of the optical encoder measuring scale.

It was shown that the measurements by the easy-to-use optical encoder can be employed to correct the first two types of positioning errors. In addition, the correction allowed us comprehensive elimination of the influence of scratches, damages, and other mechanical defects of the translation stage screw on the measured position. To correct the error of the initial position and increase the signal-to-noise ratio, it was proposed to use software correction methods.

Author Contributions: Conceptualization, P.C. and A.M.; Methodology, P.C. and A.M.; Software, F.M.; Validation, A.M., F.M., and L.M.; Formal analysis, A.M., F.M., and L.M.; Investigation, F.M. and N.N.; Resources, N.N.; Data curation, F.M.; Writing—original draft preparation, F.M. and A.M.; Writing—review and editing, A.M., P.C., and N.N.; Visualization, F.M. and A.M.; Supervision, P.C.; Project administration, P.C.; Funding acquisition, P.C.

Funding: This research was funded by the Russian Science Foundation (RSF), grant number 17-12-01418.

Acknowledgments: The authors would like to thank Sergey Kuznetsov for providing bandpass terahertz filters for 224, 264, 312, and 376 GHz.

Conflicts of Interest: The authors declare no conflict of interest.

References

1. Jepsen, P.U.; Cooke, D.G.; Koch, M. Terahertz spectroscopy and imaging - Modern techniques and applications. *Laser Photonics Rev.* **2011**, *5*, 124–166. [CrossRef]
2. Baxter, J.B.; Guglietta, G.W. Terahertz spectroscopy. *Anal. Chem.* **2011**, *83*, 4342–4368. [CrossRef] [PubMed]
3. Antsygin, V.; Mamrashev, A.; Nikolaev, N.; Potaturkin, O.; Bekker, T.; Solntsev, V. Optical properties of borate crystals in terahertz region. *Opt. Commun.* **2013**, *309*, 333–337. [CrossRef]
4. Mamrashev, A.; Nikolaev, N.; Antsygin, V.; Andreev, Y.; Lanskii, G.; Meshalkin, A. Optical Properties of KTP Crystals and Their Potential for Terahertz Generation. *Cryst.* **2018**, *8*, 310. [CrossRef]
5. Mamrashev, A.A.; Maximov, L.V.; Nikolaev, N.A.; Chapovsky, P.L. Detection of Nuclear Spin Isomers of Water Molecules by Terahertz Time-Domain Spectroscopy. *IEEE Trans. Terahertz Sci. Technol.* **2018**, *8*, 13–18. [CrossRef]
6. Xie, L.; Yao, Y.; Ying, Y. The Application of Terahertz Spectroscopy to Protein Detection: A Review. *Appl. Spectrosc. Rev.* **2014**, *49*, 448–461. [CrossRef]
7. Ulbricht, R.; Hendry, E.; Shan, J.; Heinz, T.F.; Bonn, M. Carrier dynamics in semiconductors studied with time-resolved terahertz spectroscopy. *Rev. Mod. Phys.* **2011**, *83*, 543–586. [CrossRef]
8. Lloyd-Hughes, J.; Jeon, T.-I. A Review of the Terahertz Conductivity of Bulk and Nano-Materials. *J. Infrared Millimeter Terahertz Waves* **2012**, *33*, 871–925. [CrossRef]
9. Duvillaret, L.; Garet, F.; Coutaz, J.-L. Influence of noise on the characterization of materials by terahertz time-domain spectroscopy. *J. Opt. Soc. Am. B* **2000**, *17*, 452. [CrossRef]
10. Withayachumnankul, W.; Fischer, B.M.; Lin, H.; Abbott, D. Uncertainty in terahertz time-domain spectroscopy measurement. *J. Opt. Soc. Am. B* **2008**, *25*, 1059. [CrossRef]

11. Krüger, M.; Funkner, S.; Bründermann, E.; Havenith, M. Uncertainty and ambiguity in terahertz parameter extraction and data analysis. *J. Infrared, Millimeter Terahertz Waves* **2011**, *32*, 699–715. [CrossRef]
12. Naftaly, M. Metrology issues and solutions in THz time-domain spectroscopy: Noise, errors, calibration. *IEEE Sens. J.* **2013**, *13*, 8–17. [CrossRef]
13. Zaytsev, K.; Gavdush, A.A.; Karasik, V.E.; Alekhnovich, V.I.; Nosov, P.A.; Lazarev, V.A.; Reshetov, I.V.; Yurchenko, S.O. Accuracy of sample material parameters reconstruction using terahertz pulsed spectroscopy. *J. Appl. Phys.* **2014**, *115*, 193105. [CrossRef]
14. Cohen, N.; Handley, J.W.; Boyle, R.D.; Braunstein, S.L.; Berry, E. EXPERIMENTAL SIGNATURE OF REGISTRATION NOISE IN PULSED TERAHERTZ SYSTEMS. *Fluct. Noise Lett.* **2006**, *6*, L77–L84. [CrossRef]
15. Soltani, A.; Probst, T.; Busch, S.F.; Schwerdtfeger, M.; Castro-Camus, E.; Koch, M. Error from Delay Drift in Terahertz Attenuated Total Reflection Spectroscopy. *J. Infrared Millimeter Terahertz Waves* **2014**, *35*, 468–477. [CrossRef]
16. Jahn, D.; Lippert, S.; Bisi, M.; Oberto, L.; Balzer, J.C.; Koch, M. On the Influence of Delay Line Uncertainty in THz Time-Domain Spectroscopy. *J. Infrared Millimeter Terahertz Waves* **2016**, *37*, 605–613. [CrossRef]
17. Rehn, A.; Jahn, D.; Balzer, J.C.; Koch, M. Periodic sampling errors in terahertz time-domain measurements. *Opt. Express* **2017**, *25*, 6712. [CrossRef]
18. Kim, Y.; Yee, D.-S. High-speed terahertz time-domain spectroscopy based on electronically controlled optical sampling. *Opt. Lett.* **2010**, *35*, 3715–3717. [CrossRef]
19. Dietz, R.J.B.; Vieweg, N.; Puppe, T.; Zach, A.; Globisch, B.; Göbel, T.; Leisching, P.; Schell, M. All fiber-coupled THz-TDS system with kHz measurement rate based on electronically controlled optical sampling. *Opt. Lett.* **2014**, *39*, 6482–6485. [CrossRef]
20. Humphreys, D.A.; Naftaly, M.; Molloy, J.F. Effect of time-delay errors on THz spectroscopy dynamic range. In Proceedings of the 2014 39th International Conference on Infrared, Millimeter, and Terahertz waves (IRMMW-THz), Tucson, AZ, USA, 14–19 September 2014.
21. Coakley, K.J.; Hale, P. Alignment of noisy signals. *IEEE Trans. Instrum. Meas.* **2001**, *50*, 141–149. [CrossRef]
22. Molter, D.; Trierweiler, M.; Ellrich, F.; Jonuscheit, J.; Von Freymann, G. Interferometry-aided terahertz time-domain spectroscopy. *Opt. Express* **2017**, *25*, 7547. [CrossRef] [PubMed]
23. Gordon, I.; Rothman, L.; Hill, C.; Kochanov, R.; Tan, Y.; Bernath, P.; Birk, M.; Boudon, V.; Campargue, A.; Chance, K.; et al. The HITRAN2016 molecular spectroscopic database. *J. Quant. Spectrosc. Radiat. Transf.* **2017**, *203*, 3–69. [CrossRef]
24. Kuznetsov, S.A.; Astafyev, M.A.; Gelfand, A.V.; Arzhannikov, A.V. Microstructured frequency selective quasi-optical components for submillimeter-wave applications. In Proceedings of the 2014 44th European Microwave Conference, Rome, Italy, 6–9 October 2014; pp. 881–884.
25. Arzhannikov, A.V.; Burdakov, A.V.; Vyacheslavov, L.N.; Ivanov, I.A.; Ivantsivsky, M.V.; Kasatov, A.A.; Kuznetsov, S.A.; Makarov, M.A.; Mekler, K.I.; Polosatkin, S.V.; et al. Diagnostic system for studying generation of subterahertz radiation during beam-plasma interaction in the GOL-3 facility. *Plasma Phys. Rep.* **2012**, *38*, 450–459. [CrossRef]
26. Thumm, M.K.A.; Arzhannikov, A.V.; Astrelin, V.T.; Burdakov, A.V.; Ivanov, I.A.; Kalinin, P.V.; Kandaurov, I.V.; Kurkuchekov, V.V.; Kuznetsov, S.A.; Makarov, M.A.; et al. Generation of High-Power Sub-THz Waves in Magnetized Turbulent Electron Beam Plasmas. *J. Infrared Millimeter Terahertz Waves* **2014**, *35*, 81–90. [CrossRef]

Article

Design of a 335 GHz Frequency Multiplier Source Based on Two Schemes

Jin Meng [1,*], Dehai Zhang [1], Guangyu Ji [1,2], Changfei Yao [3], Changhong Jiang [1] and Siyu Liu [1,2]

1 Key Laboratory of Microwave Remote Sensing, National Space Science Center, Chinese Academy of Sciences, Beijing 100190, China
2 School of Computer Science, University of Chinese Academy of Sciences, Beijing 100049, China
3 Laboratory of Microwave, School of Electronic and Information Engineering, Nanjing University of Information Science and Technology, Nanjing 210044, China
* Correspondence: mengjin@mirslab.cn

Received: 3 July 2019; Accepted: 25 August 2019; Published: 28 August 2019

Abstract: Based on a W-band high-power source, two schemes are proposed to realize a 335 GHz frequency multiplier source. The first scheme involves producing a 335 GHz signal with a two-stage doubler. The first doubler adopts two-way power-combined technology and the second stage is a 335 GHz doubler using a balanced circuit to suppress the odd harmonics. The measured output power was about 17.9 and 1.5 dBm at 167.5 and 335 GHz, respectively. The other scheme involves producing a 335 GHz signal with a single-stage quadrupler built on 50 μm thick quartz circuit adopting an unbalanced structure. The advantage of the unbalanced structure is that it can provide bias to the diodes without an on-chip capacitor, which is hard to realize with discrete devices. The measured output power was about 5.8 dBm at 337 GHz when driven with 22.9 dBm. Such 335 GHz frequency multiplier sources are widely used in terahertz imaging, radiometers, and so on.

Keywords: cascaded doubler; quadrupler; Schottky varactor; hybrid integrated circuit

---

1. Introduction

In recent decades, terahertz technology has been used for a variety of applications such as radio astronomy, remote sensing of the Earth's atmosphere, radar imaging, etc. [1–5]. Furthermore, advances in terahertz sources and detectors have facilitated the development of these terahertz applications. As for terahertz sources, the methods mainly include the extension of microwave electronics towards high frequencies on one side and the development of photonic devices from the optical region towards low frequencies. In general, the frequency multiplier chain is the typical electronic method that can work at room temperature.

Some leading overseas research institutes such as the Jet Propulsion Laboratory (JPL) have been able to realize solid-state frequency multiplier sources above 1 THz, which can produce tens of microwatts of power [6,7]. Furthermore, several competing technologies have been proposed in the semiconductor frequency multiplier field. In comparison, domestic research on frequency multipliers has mainly focused on the hybrid integrated circuits with discrete Schottky diodes, with an operating frequency around 200 GHz [8–10].

Based on a W-band high-power source, detailed in our former research, two schemes are proposed to design a 335 GHz multiplier. The first solution consists of a W-band source, two cascaded 167.5 GHz doublers, and a 335 GHz doubler. Another solution is to replace the two-stage doubler with a single-stage quadrupler. The details of the modules mentioned above are discussed further in this paper.

2. General Scheme

Increasing the efficiency and power capability of the frequency multiplier is the common way to obtain high output power [11–13]. For the diodes used in this paper, the safe input power was 24 dBm for a single multiplier. Therefore, power-combined technology was adopted to increase the effective input power. Figure 1 shows the diagram of a 335 GHz solid-state source with different schemes.

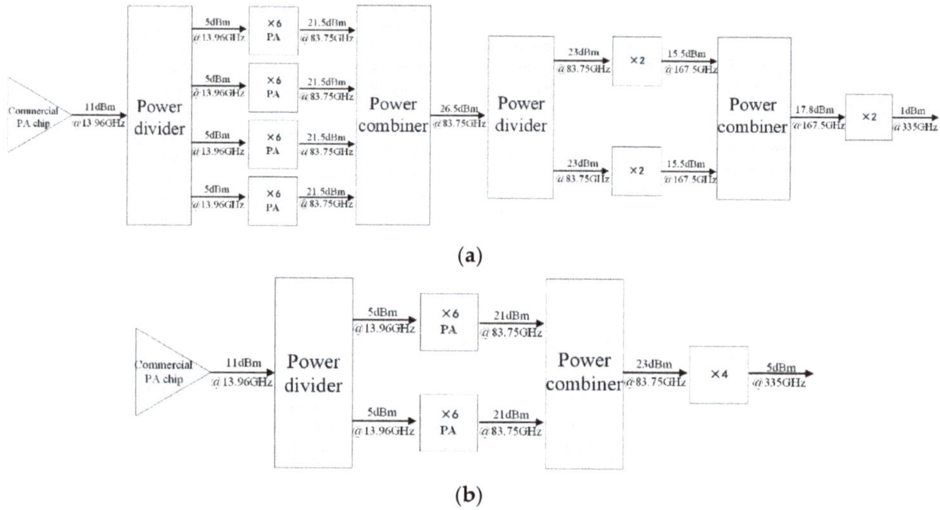

Figure 1. (**a**) Block diagram of the 335 GHz frequency multiplier based on a two-stage doubler. (**b**) Block diagram of the 335 GHz frequency multiplier based on a single-stage quadrupler.

A frequency multiplier of order N converts the input sinusoidal signal of frequency F_1 and power P_1 to an output sinusoidal signal of frequency $F_N = N \times F_1$ and power P_N. Hence, the conversion efficiency of the frequency multiplier is defined as the ratio of P_N to P_1. For a chain ($\times N_1 \times N_2$) of two cascaded multipliers of respective order N_1 and N_2, the conversion efficiency of the chain is η (N_1, N_2). A high-order frequency multiplier of order $N_3 = N_1 \times N_2$ usually has a conversion efficiency η (N_3) < η (N_1, N_2) and η (N_1, N_2) = η (N_1) × η (N_2). Therefore, in theory, the efficiency of scheme (a) in Figure 1 is higher than that of scheme (b).

Furthermore, the relation of η (N_1, N_2) = $\eta(N_1) \times \eta(N_2)$ is valid only when there is no reflected power by the second multiplier. Besides, the mismatch of the interface between the cascaded multipliers caused by dimension error and assembly error could lead to a loss of power. Consequently, the efficiency of the chain $\times N_1 \times N_2$ is not necessarily more than that of the chain $\times N_3$ in practical application.

3. Basic Principle of Schottky Diode

The Schottky barrier diode is a two-port device that is important to terahertz frequency multipliers. The diode can be divided into two modes based on operating principle: varistor and varactor. In general, the structure of a Schottky varactor is qualitatively the same as that of a varistor diode. However, the epitaxial layer of the varactor is thicker than that of the varistor, which could increase the breakdown voltage to maximize capacitance variation [14]. Figure 2 shows the cross-sectional view of the Schottky diode. The varistor makes use of a nonlinear resistance characteristic for the mixer, and the varactor is used for frequency multipliers by using a nonlinear capacitance characteristic. The upper part of Figure 2 shows the nonlinear curves of the varistor and varactor.

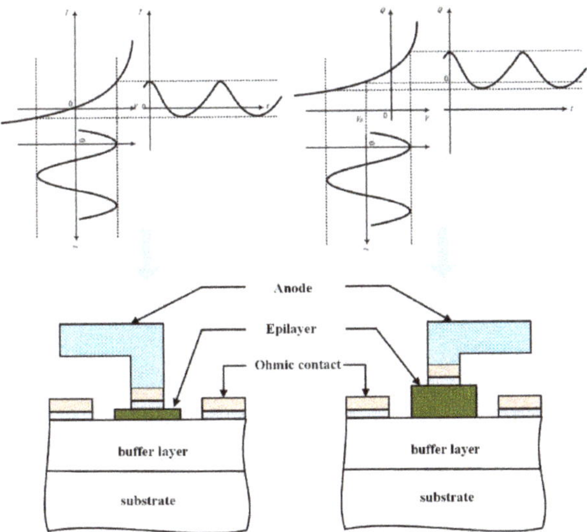

Figure 2. Structures and operating principles of the Schottky varistor and varactor.

The diode consists of intrinsic and parasitic parameters. When the frequency increases to the terahertz range, the parasitic parameters of the diode cell caused by its physical structure play an important role in affecting the performance of the frequency multiplier. Hence, the electromagnetic field around the diodes is calculated with full-wave simulation software. As for intrinsic parameters, these primarily include series resistance, zero bias junction capacitance, barrier voltage, and ideal factor. These parameters can be obtained from the IV (Intensity and voltage) or CV (Capacitance and voltage) curve.

4. Design of 335 GHz Source Based on Two-Stage Doubler

The three-dimensional model of a 335 GHz source with cascaded balanced frequency doublers is presented in Figure 3. As shown in the graph, the frequency multiplier is a split-block waveguide design, and a suspended microstrip circuit based on a 50 μm thick quartz substrate is mounted in the channel between the input and output waveguide. The varactor chips are mounted on the suspended microstrip circuit with silver epoxy.

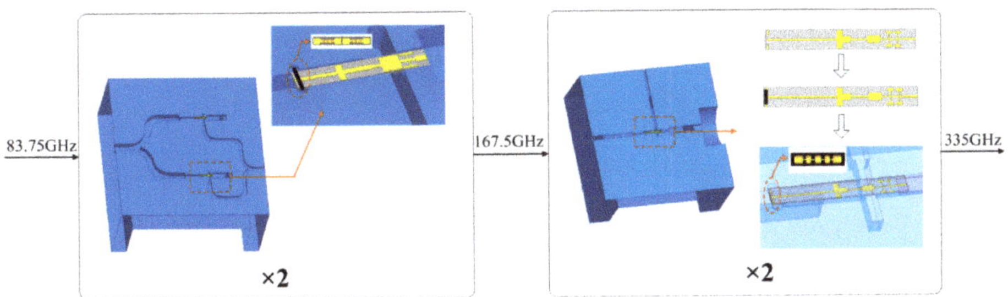

Figure 3. The three-dimensional model of the 335 GHz source based on a two-stage doubler.

4.1. W-Band High-Power Source

The W-band power source driven by the 167 GHz doubler, which mainly includes a sextupler, a power amplifier, and a four-way power-combining module. The composition block diagram is shown in Figure 4. The sextupler employs a commercially available GaAs MMIC chip HMC1110 fabricated by Analog Devices Company (Norwood, MA, USA), and the power amplifier uses an MMIC chip MAAP-011106 fabricated by M/A-COM Technology Solutions Inc (Lowell, MA, USA). Finally, the measured results indicate that the output power is more than 25 dBm at 81–86 GHz, and that the output power is about 27.5 dBm at 83 GHz when driven by 3 dBm of input power.

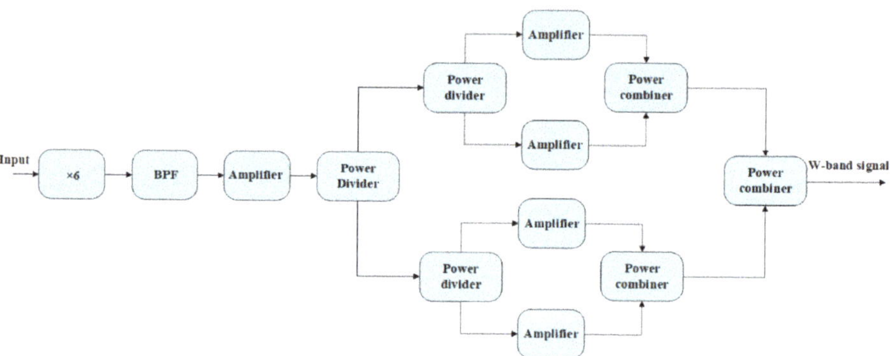

Figure 4. The block diagram of the W-band source.

4.2. Two-Way Power-Combined 167 GHz Frequency Multiplying Source

Considering the technical requirements and cost, the 167 GHz high-power frequency multiplying source adopts a two-way power-combined scheme. As depicted in Figure 5, the power source includes a power divider/combiner and two identical doublers.

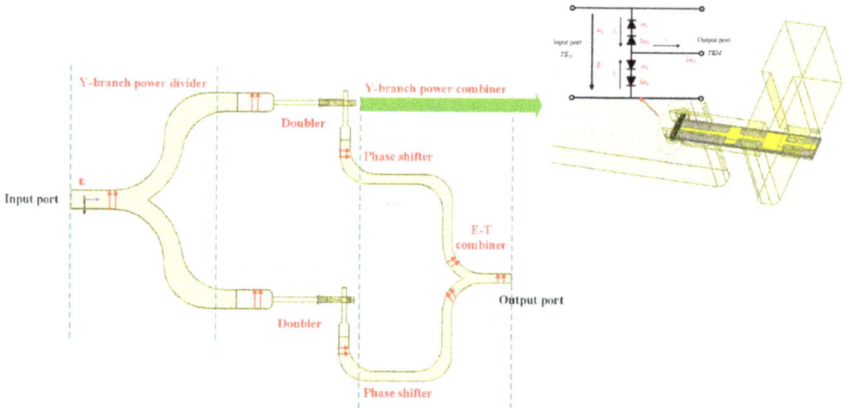

Figure 5. The structure of the 167 GHz doubler based on the two-way power-combined technology.

The power divider used in the 167 GHz high-power frequency multiplying source is a Y-type waveguide divider and the phase difference between the two output signals is zero. Furthermore, the second harmonic produced by the doublers has the same phase, and can be combined by using a Y-type waveguide combiner at the output ports. Figure 5 shows the phase relationship of each part in the two branches using the red arrow. Actually, the Y-type power combiner can be regarded as a combination of two-phase shifters and an E–T-type combiner, and the function of the phase shifter is to change the phase of the two-way signal from the same direction to the reverse direction for the T-type combiner.

Generally, the frequency doubler is designed to convert a pump microwave signal to its second harmonic based on the nonlinear voltage-dependence of the diode junction capacitance of the Schottky varactor. To suppress the odd harmonics, the diode array adopted has an anti-series type configuration. The incident signal with the dominant mode of the input rectangular waveguide (TE10) feeds the anti-series diode array. In contrast, the second harmonic would propagate along the suspended microstrip line in an unbalanced mode (TEM). In the 167 GHz doubler design, a 5VA40-13 diode chip provided by Advanced Compound Semiconductor Technologies (Hanau, Germany) was selected, which comprises a linear array of three Schottky junctions. The dimension of the chip is 240 × 60 μm (length and width, respectively) and the semi-insulating GaAs substrate is 35 μm thick.

The equivalent circuit of the balanced doubler is described at the top right of Figure 5. Based on the IV characteristic of the Schottky diode, the output current i can be expressed as [15]:

$$i = i_1 + i_2 = -i_s(e^{-\alpha V_{in}} - 1) - i_s(e^{\alpha V_{in}} - 1) = -2i_s[\cosh(\alpha V_{in}) - 1] \quad (1)$$

where i_s represents the reverse saturation current and V_{in} represents the junction voltage across the Schottky contact. By using Fourier expansion, Formula (1) is decomposed as follows:

$$i = i_s[2I_0(\alpha V_{in}) - 2] + 4i_s[I_2(\alpha V_{in})\cos(2\omega_0 t) + I_4(\alpha V_{in})\cos(4\omega_0 t) + \cdots] \quad (2)$$

where I_n (αV_{in}) is the Bessel function of the first kind. Similar to the computational method of the output current, the current in the loop can be expressed as:

$$i_{loop} = i_1 - i_2 = 4i_s[I_1(\alpha V_{in})\cos(\omega_0 t) + I_3(\alpha V_{in})\cos(3\omega_0 t) + \cdots]. \quad (3)$$

From the calculated results, it can be seen that the odd harmonics are suppressed in the output circuit, and thus the second harmonic can be obtained at the output waveguide by using the matching circuit.

4.3. 335 GHz Doubler Based on Discrete Schottky Varactor

Considering the rise of working frequency, the varactor used in the 335 GHz doubler requires a smaller zero bias junction capacitance, and therefore obtains higher cut-off frequency. At the same time, the dimension of the diode chip must be reduced to match the width of the waveguide channel. Finally, the diode chip 137C from Virginia Diodes Inc (Charlottesville, VA, USA). with four anodes in anti-series configuration was applied in the design. To improve the performance of RF ground, the flip-chip mounted method was adopted. The diode chip was glued on the ground points, which are two gold belts on either side of the quartz substrate.

The design process of the 335 GHz balanced doubler is shown in Figure 6. First of all, the impedances of the diode at fundamental and second harmonic are optimized by using source and load-pull under ideal conditions. The diode optimum impedance was found to be Z_{source} = 33 – j37 Ω and Z_{load} = 16 – j23 Ω. To improve the accuracy of simulation, the field-circuit method is applied in the design process [16,17]. Hence, the doubler is divided into two parts: a linear network, which is analyzed using the finite element method in consideration of the parasitic effects, and the nonlinear behavior of the varactor solved by the harmonic balance method. To reduce the complexity of the problem, the linear part is broken up into three sections: input transition at fundamental and second frequency, output transition, and matching circuit. The signal is coupled through a waveguide-microstrip structure, and the locations of Schottky diodes junction are inserted based on the port impedances (at fundamental frequency) acquired from step 1. Generally, the length of the reduced-height waveguide and location of the input back-short are optimized to achieve a small return loss in the input port. The second harmonic passes through the region between the diodes and input back-short and then is coupled into the output line by matching circuit. Another probe located in the output circuit couples the second harmonic to the standard output waveguide. The abovementioned design process refers to steps 2–5.

In the next step, the generated SNP files are imported to the Advanced Design System (ADS) circuit and the characteristic of the diode is added in the nonlinear circuit. Furthermore, the optimization procedure is achieved based on harmonic balance analysis. Finally, the three-dimensional model of the doubler is built according to the optimized results and the calculated S parameter of the complete circuit is exported to the ADS. Now, the doubler model is regarded as a 7-port network. The simulated result of a doubler working at 335 GHz is described at the right side of Figure 6. It clear that the odd harmonics are suppressed in simulation, and that the results coincide with those obtained by Equations (1)–(3).

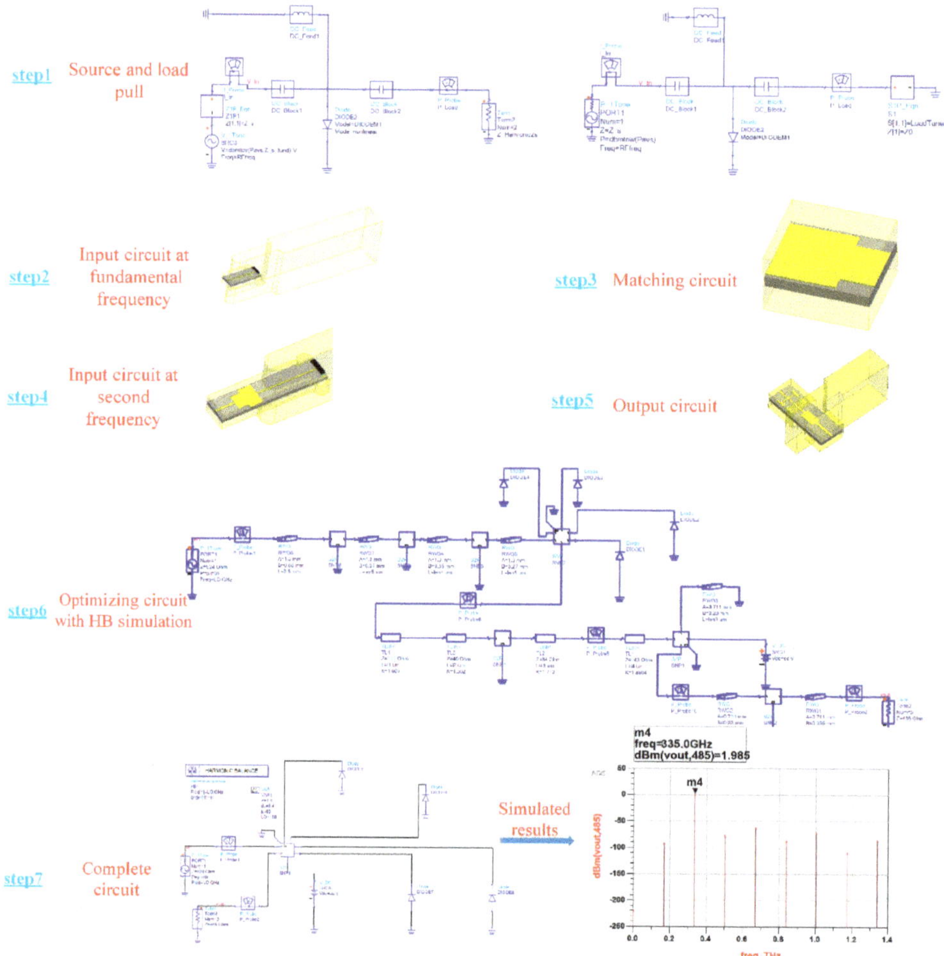

Figure 6. Design process of a 335 GHz balanced doubler with the field-circuit method.

5. Design of 335 GHz Source Based on Single-Stage Quadrupler

If varactor currents are allowed only at the input and output frequencies, a Schottky diode with ideal CV characteristics cannot generate harmonics higher than the second harmonic. To generate higher harmonics, it is necessary to allow idler currents to flow in the varactor, which could be produced by frequency doubling or frequency mixing.

For a quadrupler, one way to obtain higher harmonics is by doubling and then producing a fourth harmonic by doubling again. Another way is by mixing the second harmonic idler with the fundamental to produce the third harmonic idler first, and by continuing to mix to produce the fourth harmonic output. Actually, the high-order multipliers are most efficient when idler circuits are provided at all idler frequencies. Therefore, the use of idlers could increase the output power and efficiency of reactive frequency multipliers.

The quadrupler has a suspended microstrip circuit based on a 50 μm thick quartz substrate mounted in the channel with silver epoxy. The input and output ports of the quadrupler are standard full-height WR-12 and WR-2.8 waveguides with waveguide dimensions of 3.1 × 1.65 mm^2 and 0.71 × 0.355 mm^2, respectively. As described in Figure 7, a compact suspended microstrip resonator (CSMR) low-pass filter follows an input conversion structure (the simulated return loss is below −20 dB from 75 to 90 GHz), and hence the fourth harmonic produced by varactors could prevent leaking from the input port. Compared with the step impedance filter, the CSMR filter has a compacted structure and wide stop band [18]. The simulated result of the CSMR filter in the band of 30–350 GHz are shown in Figure 7. It can be seen from the graph that the insert loss is lower than 0.3 dB in the pass band, while the side rejection is better than 20 dB from 160 to 350 GHz. Finally, another probe located in the output circuit couples the fourth harmonic to the standard output waveguide, and the simulated return loss is better than 15 dB in the frequency range of 320–360 GHz. All passive networks, such as the low-pass filter, input and output waveguide-to-microstrip transition, and diode passive part, are analyzed by EM simulators. When the sub-circuits are optimized, the complete quadrupler circuit is simulated. The nine port S-parameters of this simulation are extracted and then combined with a nonlinear diode to model the multiply efficiency in the circuit simulator. This process is usually repeated for the further optimization of the quadrupler multiply efficiency.

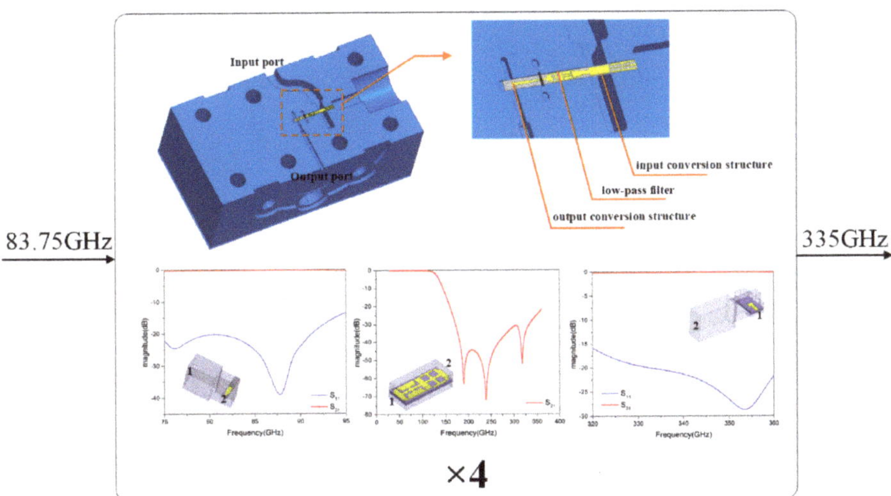

Figure 7. Three-dimensional model of the 335 GHz source based on a single-stage quadrupler.

6. Measurements and Discussion

The block diagram of the measurement setup is illustrated in Figure 8. An Agilent analog signal generator E8257D (Santa Clara, CA, USA) is followed by the W-band power source to generate the signal in the 81–86 GHz band. The output power of the frequency multipliers is measured by a PM4 power meter (Charlottesville, VA, USA). Moreover, a Sub-Miniature-A (SMA) (type KFD55(Xi'an, China)) is connected to the main transmission circuit using gold wire bonding and an external sliding rheostat connected to the SMA port so as to bias the varactor.

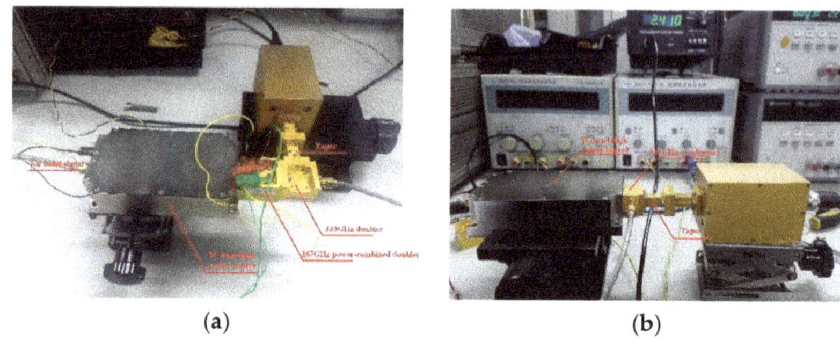

Figure 8. (**a**) Photo of the assembled 335 GHz source with two cascaded doublers. (**b**) Photo of the assembled 335 GHz source with a single-stage quadrupler.

Figure 9a shows the measured results of the W-band source. The measured output power is more than 310 mW from 81 GHz to 86 GHz, and the maximum power is about 560 mW at 83 GHz. In the measurement of the quadrupler, the input power is about 200 mW to make sure the diodes work safely, and hence an attenuator is added. The measured output power of the 167 GHz power-combined doubler is shown in Figure 9b. It was found that the power is more than 45 mW at 164–172 GHz, and the highest output power is 62 mW at 167.5 GHz.

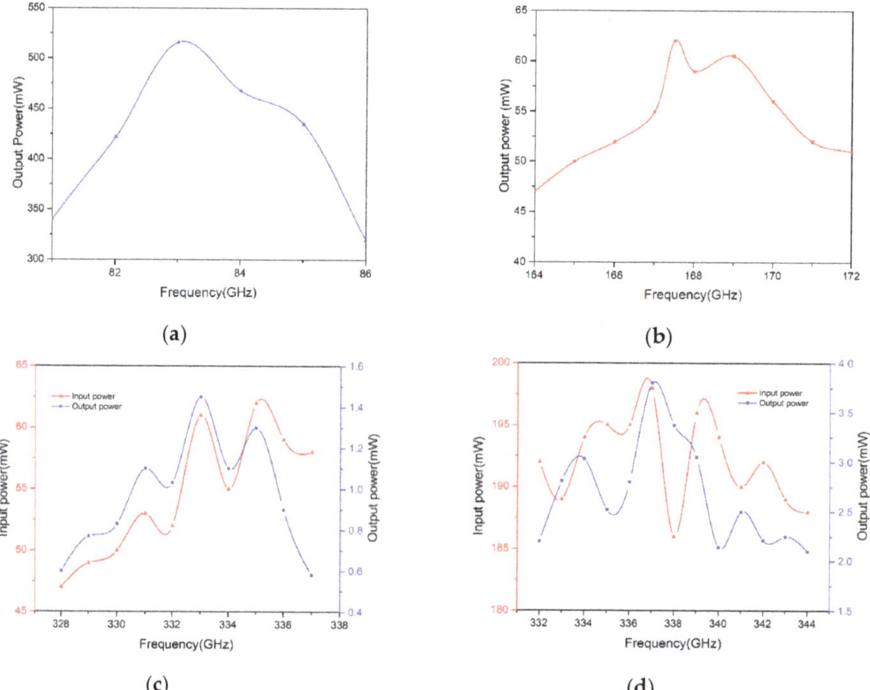

Figure 9. (**a**) Measured output power of the W-band source. (**b**) Measured output power of the 167 GHz high-power source. (**c**) Measured input and output power of the 335 GHz source based on a two-stage doubler. (**d**) Measured input and output power of the 335 GHz source based on a single-stage quadrupler.

The measured results of the 335 GHz source based on two cascaded doublers are shown in Figure 9c. The measured output power of the 335 GHz doubler is more than 0.5 mW at 328–337 GHz and the maximum output power is about 1.4 mW at 333 GHz. The relation of the single-stage quadrupler 335 GHz source output versus pumping power is described in Figure 9d. The measured typical output power is 2.5 mW at 332–344 GHz, and the highest measured output power of 3.8 mW is measured at 337 GHz with an input power of 198 mW.

Tables 1 and 2 illustrate a comparison of some reported multipliers. A recent development of terahertz solid circuits in China caused a regression in advanced semiconductor technology such as Schottky diode technology, membrane technology, transferred substrate technology, and so on. The result is that an integrated circuit is difficult to realize and the design of the frequency multiplier needs to use a discrete circuit with a higher loss transmission line. The performance of the frequency multipliers presented in this paper reached the same level as that achieved by research institutions abroad, and has a leading position at home. Furthermore, the design using discrete diodes is easy to realize and the cost is relatively low.

Table 1. Performance comparison of the doubler around 170 GHz.

References	Technology	Frequency (GHz)	Max Output Power (mW)	Typical Efficiency
[19] from VDI (Virginia Diodes, Inc.)	×2 integrated	110–170	24	8%
[20] from NUIST (Nanjing University of Information Science and Technology)	×2 discrete	176–196	40	12%
This paper	×2 discrete	164–172	62	20%

Table 2. Performance comparison of the frequency multipliers above 300 GHz.

References	Technology	Frequency (GHz)	Max Output Power (mW)	Typical Efficiency
[19] from VDI	×2 integrated	220–330	3	6%
[21] from RAL (Rutherford Appleton Laboratory)	×2 discrete	330–336	1.2	2%
[22] from UESTC (University of Electronic Science and Technology of China)	×3 discrete	320–342	0.149	0.3%
This paper	×2 discrete	328–337	1.4	2%
	×4 discrete	332–344	3.8	1.5%

To ensure that the 167 GHz doubler is working safely, the input power produced by the W-band source is controlled below 280 mW and the typical output power of the doubler is, accordingly, 55 mW. Hence the typical efficiency of the 167 GHz doubler is about 20%. The typical efficiency of the 335 GHz doubler is 2% with typical input and output power values of 55 and 1.1 mW, respectively. To summarize, the efficiency of the frequency multiplier chain is about 0.4%. In contrast, the typical input and output power values of a single-stage quadrupler are 190 mW and 2.8 mW, and said quadrupler has a higher efficiency of about 1.5%. However, it is difficult to say whether it is better to realize a high-order multiplier via a single-stage process or by a cascade of two or more low-order multipliers. The decision could be made in accordance with specific conditions.

7. Conclusions

A solid-state frequency multiplier chain based on two schemes has been designed and tested in this paper. For the first option, the measured highest output power was about 1.4 mW at 333 GHz and more than 0.5 mW at 328–337 GHz. For the second option, the measured typical output power was 2.5 mW at 332–344 GHz, and the highest measured output power was 3.8 mW at 337 GHz. The research content provides the means to generate a terahertz signal above 300 GHz. Our future work will aim at the design of a G-band, higher output power source, and at the increased efficiency of the frequency multiplier working in the sub-millimeter region.

Author Contributions: Conceptualization, methodology, software, and writing, J.M., D.Z., and G.J.; formal analysis, C.Y. and C.J.; visualization, J.M. and S.L.

Funding: This research received no external funding.

Conflicts of Interest: The authors declare no conflict of interest.

References

1. Chattopadhyay, G. Terahertz antennas and systems for space borne platforms. In Proceedings of the Fourth European Conference on Antennas and Propagation (IEEE), Barcelona, Spain, 12–16 April 2010.
2. Arora, A.; Luong, T.Q.; Krüger, M.; Kim, Y.J.; Nam, C.H.; Manz, A.; Havenith, M. Terahertz-time domain spectroscopy for the detection of PCR amplified DNA in aqueous solution. *Analyst* **2012**, *137*, 575–579. [CrossRef] [PubMed]
3. Waters, J.W.; Froidevaux, L.; Harwood, R.S.; Jarnot, R.F.; Pickett, H.M.; Read, W.G.; Siegel, P.H.; Cofield, R.E.; Filipiak, M.J.; Holden, J.R.; et al. The earth observing system microwave limb sounder (EOS MLS) on the aura satellite. *IEEE Trans. Geosci. Remote Sens.* **2006**, *44*, 1075–1092. [CrossRef]
4. Maestrini, A.; Ward, J.; Chattopadhyay, G.; Schlecht, E.; Mehdi, I. Terahertz Sources Based on Frequency Multiplication and Their Applications. *Frequenz* **2008**, *62*, 118–122. [CrossRef]
5. Moyna, B.P.; Charlton, J.E.; Lee, C.; Parker, R.J.; Oldfield, M.M.; Matheson, D.N.; Peter, M.; Kangas, V. Design of a sub-millimetre wave airborne demonstrator for observations of precipitation and ice clouds. In Proceedings of the Antennas and Propagation Society International Symposium, Charleston, SC, USA, 1–5 June 2009.
6. Imran, M.; Bertrand, T.; Robert, L.; Alain, M.; John, W.; Erich, S.; John, G.; Choonsup, L.; Goutam, C.; Frank, M. High-power local oscillator sources for 1–2 THz. *Int. Soc. Opt. Eng.* **2010**, *7741*. [CrossRef]
7. Chattopadhyay, G.; Schlecht, E.; Ward, J.S.; Gill, J.J.; Javadi, H.H.; Maiwald, F.; Mehdi, I. An All-Solid-State Broad-Band Frequency Multiplier Chain at 1500 GHz. *IEEE Trans. Microw. Theory Tech.* **2004**, *52*, 1538–1547. [CrossRef]
8. Yao, C.; Wei, X.; Luo, Y.; Zhou, M. A 210 GHz power-combined frequency multiplying source with output power of 23.8 mW. *Frequenz* **2017**, *71*, 523–530. [CrossRef]
9. Zhang, Y.; Lu, Q.-Q.; Liu, W.; Li, L.; Xu, R.-M. Design of a 220 GHz frequency tripler based on EM model of Schottky diodes. *J. Infrared Millim. Waves* **2014**, *33*, 405–411.
10. Yao, C.F.; Zhou, M.; Luo, Y.S.; Wang, Y.G.; Xu, C.H. 150 GHz and 180 GHz fixed-tuned frequency multiplying sources with planar Schottky diodes. *J. Infrared Millim.Waves* **2013**, *32*, 102–107. [CrossRef]
11. Maestrini, A.; Ward, J.S.; Tripon-Canseliet, C.; Gill, J.J.; Lee, C.; Javadi, H.; Chattopadhyay, G.; Mehdi, I. In-phase power-combined frequency triplers at 300 GHz. *IEEE Microw. Wirel. Compon. Lett.* **2008**, *18*, 218–220. [CrossRef]
12. Siles, J.V.; Maestrini, A.; Alderman, B.; Davies, S.; Wang, H.; Treuttel, J.; Leclerc, E.; Narhi, T.; Goldstein, C. A Single-Waveguide in-Phase Power-Combined Frequency Doubler at 190 GHz. *IEEE Microw. Wirel. Compon. Lett.* **2011**, *21*, 332–334. [CrossRef]
13. Gupta, M.S. Degradation of power combining efficiency due to variability among signal sources. *IEEE Trans. Microw. Theory Tech.* **1992**, *40*, 1031–1034. [CrossRef]
14. Faber, M.T.; Chramiec, J.; Adamski, M.E. *Microwave and Millimeter-Wave Diode Frequency Multipliers*; Artech House: Boston, MA, USA; London, UK, 1995.
15. Penfield, P.; Rafuse, R.P. *Varactor Applications*; MIT Press: London, UK, 1962.
16. Thomas, B.; Treuttel, J.; Alderman, B.; Matheson, D.; Narhi, T. Application of substrate transfer to a 190 GHz frequency doubler and 380 GHz sub-harmonic mixer using MMIC foundry Schottky diodes. In Proceedings of the Millimeter and Submillimeter Detectors and Instrumentation for Astronomy IV, Marseille, France, 23–28 June 2008.
17. Meng, J.; Zhang, D.; Jiang, C. Research on the practical design method of 225 GHz tripler. *J. Infrated Millim. Waves* **2015**, *34*, 190–195.

18. Yang, X.; Zhang, B.; Fan, Y.; Zhong, F.Q.; Chen, Z. Design of improved CMRC structure used in terahertz subharmonic pumped mixer. In Proceedings of the 12th IEEE International Conference on Communication Technology, Nanjing, China, 11–14 November 2010.
19. Vadiodes. Available online: http://www.vadiodes.com (accessed on 21 August 2019).
20. Chen, Z.; Chen, X.; Cui, W.; Li, X.; Ge, J. A High-Power G-band Schottky local oscillator chain for submillimeter wave heterodyne detection. *J. Infrared Milli. Terahertz Waves* **2015**, *36*, 430–444. [CrossRef]
21. Liu, H.; Powell, J.; Viegas, C.; Cairns, A.A.; Alderman, B. A 332 GHz frequency doubler using flip-chip mounted planar Schottky diodes. In Proceedings of the Asia-Pacific Microwave Conference, Nanjing, China, 6–9 December 2015.
22. Xiao, Z. *Research on 330 GHz Multiplier Based on Planar Schottky Diode*; University of Electronic Science and Technology of China: Chengdu, China, 2017.

Article

Improvement in SNR by Adaptive Range Gates for RCS Measurements in the THz Region

Shuang Pang, Yang Zeng *, Qi Yang, Bin Deng, Hongqiang Wang and Yuliang Qin

College of Electronic Science and Technology, National University of Defense Technology, Changsha, 410073, China
* Correspondence: zengyang@nudt.edu.cn; Tel.: +86-13875897569

Received: 14 May 2019; Accepted: 15 July 2019; Published: 18 July 2019

Abstract: One of the major concerns in radar cross-section (RCS) measurements is the isolation of the target echo from unwanted spurious signals. Generally, the method of software range gate is applied to extract useful data. However, this method may not work to expectations, especially for targets with a large length-width ratio. This is because the effective target zone is dependent on the aspect angle. The implementation of conventional fixed range gates will introduce an uneven clutter signal that leads to a decline in signal-to-noise ratio. The influence of this uneven clutter signal becomes increasingly severe in the terahertz band, where the wavelength is short and the illumination power is weak. In this work, the concept of adaptive range gates was adopted to extract a target echo of higher accuracy. The dimension of the range gate was determined by the angle-dependent radial projection of the target. In order to evaluate the performance of the proposed method, both experimental measurements and numerical simulations were conducted. Noticeable improvements in the signal-to-noise ratio at certain angles were observed.

Keywords: Terahertz radar; radar cross-section; signal-to-noise ratio; adaptive range gates

---

1. Introduction

The radar cross-section (RCS) of a target is an important physical quantity that characterizes its capability of reflecting electromagnetic waves. The RCS is a physical quality of high complexity. It is related to not only the geometrical parameters and physical parameters of the target, such as the size, shape, material, and structure, but also the parameters of the incident wave, including the frequency, polarization waveform, etc. [1,2]. High-quality RCS data are strongly required for programs such as high-range-resolution profiling (HRRP) and synthetic aperture radar (SAR) imaging. Commonly, two different approaches are implemented in RCS measurements. One is to measure the RCSs of the full-size objects; the other is to measure the RCSs of scaled models and calculate the RCSs of the objects according to the law of electromagnetic similarity. These two approaches have their respective pros and cons [3,4]. Direct full-size measurements can be easily arranged for targets with moderate sizes within appropriate test sites. However, in terms of large objects, this method will become costly or even impossible owing to the requirement in large test-field and long measurement distance. In contrast, scaled measurements for large prototypes can be carried out in anechoic chambers; while this method requires radar systems with high-performance transceiver links and high operation frequencies. Therefore, obtaining RCS data in the terahertz (THz) band is of great significance for scaled measurements. In recent years, owing to the growing demand in RCS scaled measurements of large objects and producing vehicles with reduced RCS in the terahertz range, the development of THz RCS measurement systems has attracted increasing attention [5,6]. In accordance with the approach of THz wave generation and detection, various types of THz RCS measurement systems have been constructed, including THz laser-based systems, THz time-domain spectroscopy (THz-TDS)-based systems and solid-state THz systems [7–10].

It is realized that THz RCS measurements are very different from conventional RCS measurements. In THz band, the range resolution is significantly improved thanks to the large time-bandwidth product accompanied by the high carrier frequency. As a result, the contribution of different scattering centers can be isolated and analyzed. This is highly preferred in high-resolution radar imaging [11–14]. Meanwhile, due to the shortened wavelength of THz radiation, the contribution of the fine structure of the targets, such as surface roughness, can no longer be ignored [15,16]. This provides new insights regarding scattering mechanism studying. Therefore, the development of an accurate THz RCS measurement technique is in strong demand. However, with the increasing operation frequency, the transmitting power of the radar system decreases dramatically [17,18]. Besides, the influence of fluctuation in the signal source and measurement environment becomes increasingly significant. All these factors limit the precision of THz RCS measurements.

The performance of the THz RCS measurement system can be improved by upgrading the hardware and bettering the data processing technique. The former, however, is generally expensive and time-consuming. This work specifically focuses on the latter approach. Conventionally, the software range gate technique is applied in RCS measurements to filter the target echo from unwanted spurious signals and subsequently improve the signal-to-noise ratio (SNR). Range gates are generally configured with fixed dimensions that correspond to the size of the quiet zone [19]. This technique works well in microwave frequency measurements, as the range resolution is relatively low and small changes in radial projection size will not influence the range profile. However, for THz radars, the range resolution is significantly improved. Hence, an alteration in radial projection, corresponding to the aspect angle, can no longer be ignored. In this scenario, a fixed range gate will bring in the uneven contribution of the clutter signal corresponding to the aspect angle. The influence of this uneven clutter signal becomes increasingly significant in THz RCS measurements considering the nature of weak transmitting power in THz radars. As a result, the conventional range gate may not work up to expectations in THz RCS measurements, especially for targets with large length-width ratios.

To alleviate this problem, the new concept of an adaptive range gate is proposed in this work. The dimension of the adaptive range gate is adjusted according to the observation angle. This will effectively eliminate the uneven contribution of the clutter signal and improve the accuracy of RCS measurement. In the following sections, experimental measurements and numerical simulations will be carried out to evaluate the performance of our proposed method.

This paper is organized as follows. The theory of the step-frequency signal model and the concept of the adaptive range gate method are introduced in detail in Section 2. In Section 3, the THz RCS measurement system and the experiments are described. Measurement and correlated simulation results are presented and discussed in Section 4 to demonstrate the superiority of the adaptive range gate method. Conclusions are drawn in Section 5.

2. Method of the Adaptive Range Gate

2.1. The Signal Model of Step-Frequency Systems

Step-frequency (SF) signals are commonly used in RCS measurement systems because of its large bandwidth and high range resolution. In terms of the high-resolution frequency-stepped millimeter-wave (MMW) radar systems, a radar target is considered containing several strong scattering centers. By employing inverse Fourier transformation to the SF echoes, the synthetic range profile can be obtained, as well as the distribution of scattering centers [19,20].

The time-frequency signal shown in Figure 1a represents a typical transmitted SF signal. The SF signal is composed of a sequence of pulses with different frequencies separated by a fixed spectral interval. The frequency of the signal varies by step in time, while generating an even spectral power distribution, as illustrated in Figure 1b. The transmitted SF signal $S(f)$ can be described as the sum for a series of impulse functions:

$$S(f) = \sum_{i=0}^{N-1} \delta[f - (f_0 + i\Delta f)], i = 0, 1, \cdots, N-1 \tag{1}$$

where N is the number of impulses, f_0 is the initial frequency, and Δf is the spectral step size. In the receiving end, the received signal corresponding to the ith frequency point at time t can be expressed as $y_i(t)$:

$$y_i(t) = exp[-j2\pi(f_0 + i\Delta f) \cdot t], i = 0, 1, \cdots, N-1 \tag{2}$$

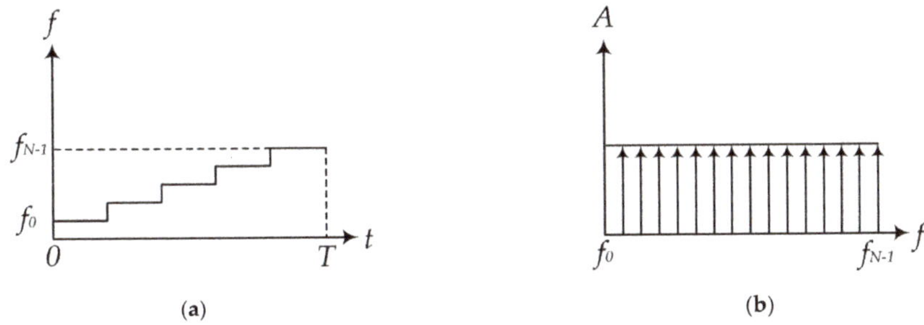

Figure 1. Evolution of a typical step-frequency (SF) signal composed of N pulses. (**a**) The frequency of the pulses increases by step in time; (**b**) The model of signals generated by the programmable network analyzer (PNA).

Considering a scattering center at range cell R, the echo signal corresponding to the ith pulse is obtained by replacing t with $2R/c$:

$$y_i(R) = exp[-j2\pi(f_0 + i\Delta f) \cdot 2R/c], i = 0, 1, \cdots, N-1 \tag{3}$$

where c is the speed of light.

The received echo can be considered as the sum of the ith impulse response $y_i(R)$ from the scattering center at range cell R. The distribution of the echo signal in the discrete-time domain $H(l;R)$ can be obtained by implementing inverse Fourier transform on the received signal sequence $\Sigma y_i(R)$ ($i = 0, 1, \ldots, N-1$) as in Equation (4):

$$H(l;R) = \frac{1}{N}\sum_{i=0}^{N-1} exp[-j2\pi(f_0 + i\Delta f) \cdot 2R/c] \cdot exp\left(j\frac{2\pi}{N}li\right), l = 0, 1, \cdots, N-1 \tag{4}$$

where l denotes the index of the time cell. The modulus of $H(l;R)$ represents the distribution of power in the time domain, i.e., the range profile:

$$\left|H(l;R)\right| = \left|\frac{sin\pi(l - N\Delta f \cdot 2R/c)}{Nsin\pi(l/N - \Delta f \cdot 2R/c)}\right|, l = 0, 1, 2, \cdots, N-1 \tag{5}$$

As in Equation (5), $H(l;R)$ is in the form of a series of sinc functions [20]. The maximum of $|H(l;R)|$ is realized when $l_0 = 2N\Delta fR/c$. This corresponds to the radial distance of the scattering center $R = cl_0/2N\Delta f$. After obtaining the range profile, a range gate corresponding to the dimension of the target zone is commonly applied in order to filter echoes from thc target and suppress the clutters. This can effectively improve the quality of the signal. Hence, properly configured range gates are vital for accurate RCS measurements.

In THz RCS measurements, SF signals are transmitted and received through the transceiver links. Due to the high range resolution in the THz band, the scattering from targets in the terahertz band can be generally considered as multiple isolated scattering centers. In this sense, the distribution of the scattering centers could be obtained according to the precise synthetic range profile by implementing the above procedure. However, a problem arises when applying the range gates. For a specific object, the target zone is, in fact, dependent on the aspect angle. As a result, at certain aspect angles, a range gate of a fixed dimension will bring extra clutters to the signal, which leads to reductions in the SNR and measurement accuracy. This becomes increasingly significant in THz RCS measurements, especially for targets with a large length-width ratio, since the SNR of THz RCS measurement systems is low and THz waves are highly sensitive to alterations in distance. Addressing this problem, the adaptive range gate method is introduced in the next subsection as an improved scheme of data extraction in the time domain.

2.2. Method of the Adaptive Range Gate

Comparing to traditional range gates, the dimension of adaptive range gates varies according to the angle-dependent target zone to precisely capture the echoes from the target and improve the SNR in RCS measurements. Figure 2 illustrates the schematic of the adaptive range gate. The rectangle with sides h and d represents the target area according to its characteristic dimensions and α is the aspect angle. The projection dimension in radial direction L' and the size of the time-domain window for adaptive range gate M' can be expressed as follows:

$$L' = d|sin\alpha| + h|cos\alpha| \tag{6}$$

$$M' = \lceil L'/h \rceil \cdot M \tag{7}$$

where M denotes the range gate corresponding to h. It is obtained by analyzing the distribution of scattering centers in the range profile at α = 0. In the adaptive range gate scheme, although the initial position in the time-domain echo varies with the azimuths angle, the gate center is considered as stationary. Here, the strongest echoes in the time domain at 0° and 90° are located and the gate center is selected according to the relationship between the geometrical size and range resolution.

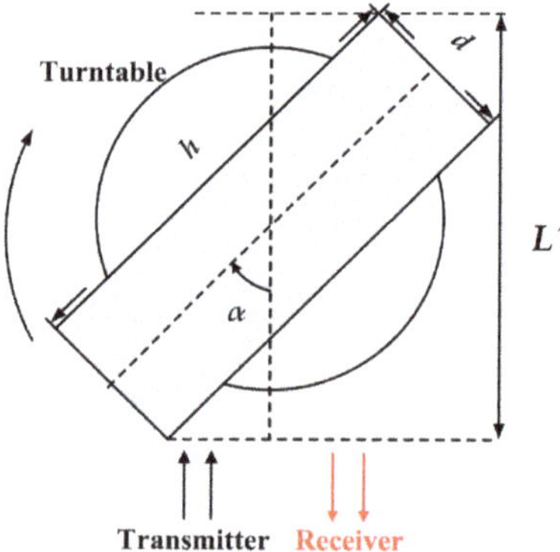

Figure 2. Schematic of the adaptive range gate.

Figure 3 highlights the difference between the fixed range gates and the adaptive range gates. The size of a fixed range gate is determined by the largest dimension of the target, while the size of the adaptive range gate is adjusted according to the projection dimension in the radial direction. As illustrated in Figure 3a, the size of adaptive range gate is constantly smaller than that of the fixed range gates (Case 2 and Case 3) with an only exception at the direction corresponding to the diagonal of the target area (Case 1). Figure 3b further provides a full-angle comparison between conventional and adaptive range gates according to Equations (6) and (7). The adaptive range gate can effectively capture the echoes from the target, while the clutter in the dashed area is eliminated.

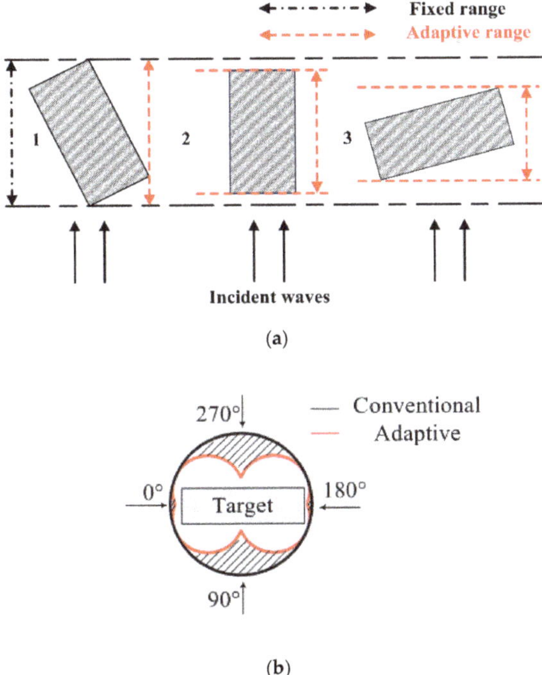

Figure 3. Comparison of two range gates at different azimuths. (**a**) Certain angles, (**b**) 0° to 360°.

3. The THz RCS Measurement System and Experiments

3.1. The Terahertz RCS Measurement System

To verify the effectiveness of the adaptive range gate method, a THz RCS measurement system was established and experiments based on a smooth metal cylinder were carried out. The measurement system consisted of the following parts: The THz radar system, the turntable with double axis and the system control center.

The THz radar system is the core of the THz RCS measurement system. The main components include the programmable network analyzer (PNA) and the frequency multiplier chains. The PNA provides the RF signal, intermediate frequency (IF) procession, as well as the data collection function. The multiplier chains contain the amplifier, doublers, mixer, and the transmitting and receiving antennas. The THz radar system can provide different carrier frequencies by changing the multipliers and antennas. In this paper, we constructed 220 GHz and 440 GHz radar systems for the RCS measurements. The two systems are similar except for the antennas and multipliers. In the 220 GHz system, the THz signals are transmitted and received by conical-horn antennas, while in the 440 GHz system, pyramidal-horn antennas are used. A schematic diagram of the radar systems and the THz multiplier chains are shown in Figures 4 and 5, respectively.

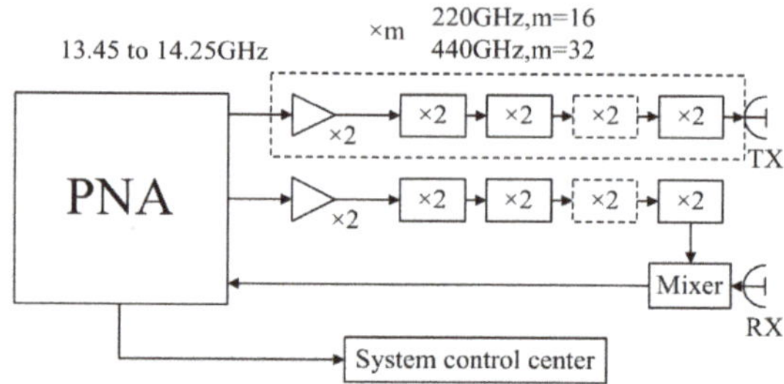

Figure 4. The schematic diagram of the radar system.

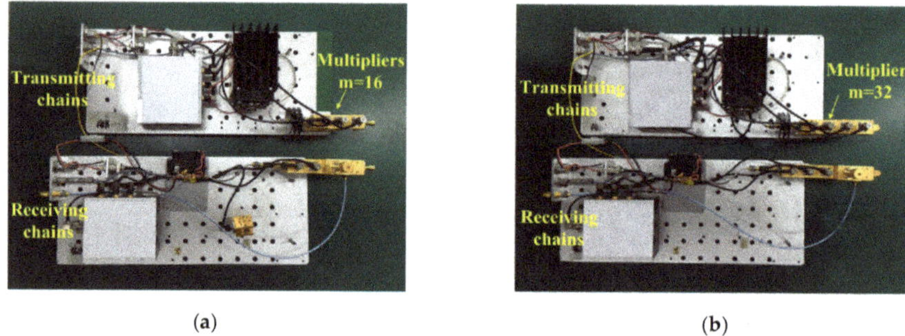

Figure 5. The THz multiplier chains. (**a**) 220 GHz, (**b**) 440 GHz.

The PNA provides a baseband signal varying between 13.45 and 14.25 GHz, and an IF signal of 960 MHz. The baseband signal is multiplied into the THz band after power amplifying. The orders of multiplication m for 220 and 440 GHz systems are 16 and 32, respectively. The available bandwidths are 12.8 GHz and 25.6 GHz, and their output powers are about 100 mW at 220 GHz and 5 mW at 440 GHz. In the receiving chain, the received THz signal is first down-converted to IF for super heterodyne reception through harmonic mixing. Then, the IF signal is down-converted to baseband and demodulated by an I/Q demodulator for A/D sampling. Finally, the collected data are transmitted to the system control center through the Ethernet for further processing.

3.2. The Experiments on a Metallic Cylinder

Experiments were carried out in a chamber equipped with absorbing material designed for THz operation. The absorption coefficient of the material at 200~500 GHz reached −36 dB, which could effectively suppress background noise and reduce the influence of multi-path interference such as the reflection wave from the wall and ground. In the experiment, the radar system was placed on a platform and the antennas were allocated side by side with a 5 cm gap between their front ends. This configuration could effectively suppress the influence of the direct leakage signal between the transmitter and the receiver. The target was placed on a foam bracket fixed on the turntable and it was 3.30 m away from the center of the antennas. The distance was much larger than the gap mentioned above, so the measurement system could be regarded as a mono-static system. In addition, the model center and the antenna center were on the same height and in the same vertical plane. With this experimental configuration, the dimension of the quiet zone was estimated as 30–40 cm in diameter, which was sufficiently large enough to cover the under-measurement target. The configuration of the measurement system and the metal cylinder are presented in Figure 6a,b.

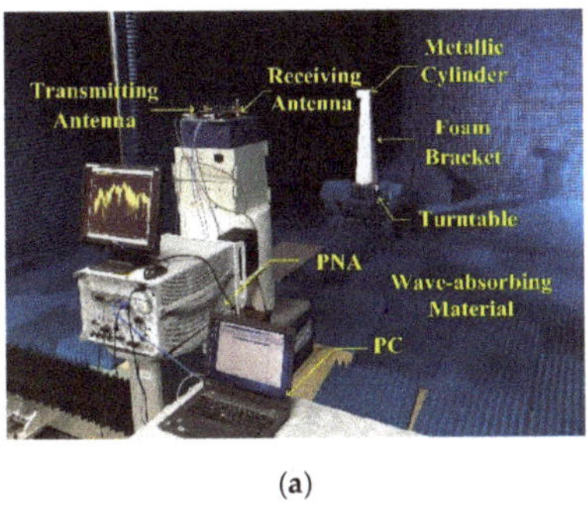

(a)

(b)

Figure 6. The THz radar cross-section (RCS) measurement scene. (**a**) The measurement system, (**b**) The cylinder.

The RCS of a metallic cylinder (of height 20 cm and radius 4 cm, shown in Figure 6b) at 220 GHz and 440 GHz were measured in the experiments. The turntable was controlled to rotate from 0° to 360° in 0.2° increments. The PNA was configured to operate at a stepped continuous wave mode so that the range profile could be obtained from Equation (5). The number of sweeping points was set at 801 for both frequency bands.

3.3. Data Processing and Calibration of Measurement Results

The data processing procedure to obtain the RCS results from the data measurement by the PNA is introduced in this subsection. This procedure mainly contained echo signal extraction from the target zone and calibration. Figure 7 presents the data proceeding process of the two range gate methods. First, for the received echo of aspect angle α, inverse Fourier transform was applied to obtain time-domain sequences. The range profile of aspect angle α was obtained according to Equation (5). Then, range gates were implemented to extract the target data from the range profile. For the adaptive range gate method, the size of the range gate was decided by the radial projection size of aspect angle α and the maximum radial projection size of the target. As for the size of the conventional range gate method, a fixed value determined by the maximum radial projection size of the target was applied. The data filtered by the range gate of aspect angle α corresponded to the echo of the scattering center. After calibration, the measured RCS of the target at aspect angle α was finally obtained.

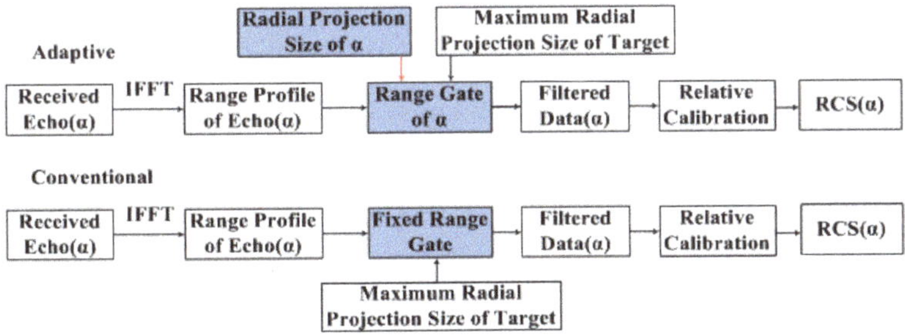

Figure 7. The flowchart of the data extraction procedure to obtain the RCS of the target by the two range gate methods.

Here in this work, the method of the relative calibration was adopted. In the experiments, the RCS of a standard object (a metallic sphere) was measured in advance under the same condition. The influence of the measurement environment was, therefore, accounted in the relative calibration procedure [19]. The calibrated RCS result of the measured target at a certain aspect angle $\sigma_t(\alpha)$ could then be obtained:

$$\sigma_t(\alpha) = 10lg\frac{|E_t(\alpha)|^2}{|E_0(\alpha)|^2} + \sigma_0(\alpha) \tag{8}$$

where E_0 (α) is the measured echo of the metallic sphere, $E_t(\alpha)$ is the measured echo of the target, and $\sigma_0(\alpha)$ is the theoretical value of the metallic sphere.

4. Experimental Results and Analysis

The measurement results are presented in this section. Figure 8a illustrates the correlation between the azimuthal angle and the position of the measured cylinder. The measured range profiles at the aspect angle of 0° and 90° in the time domain at 220 GHz and 440 GHz are shown in Figure 8b,c, respectively.

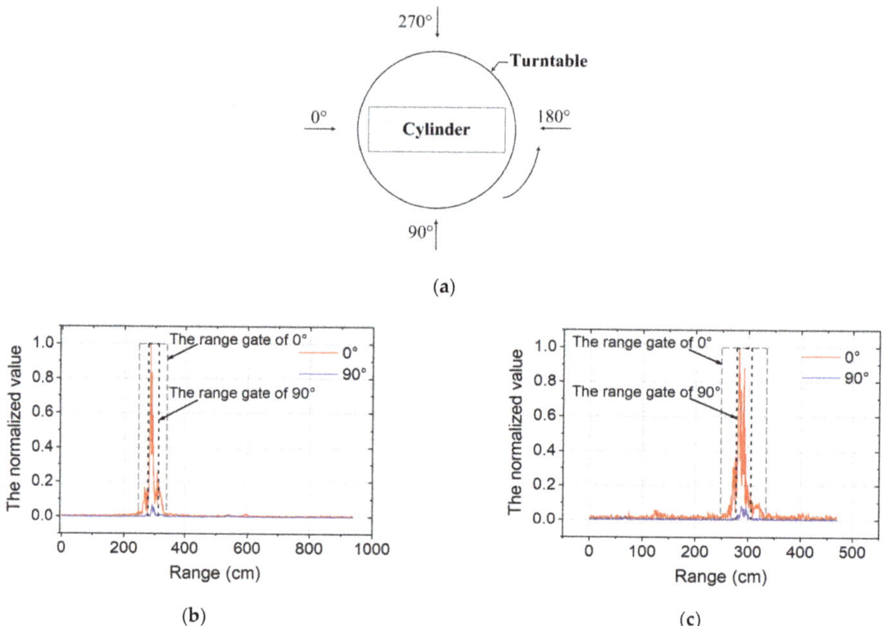

Figure 8. The azimuthal angle configuration of the measurement (**a**) and the range profiles at 0° and 90° in the time domain and the respective adaptive range gates. (**b**) 220 GHz, (**c**) 440 GHz.

As demonstrated, the distribution of the scattering zone in the range profile was dependent on the azimuthal angle. This was caused by the variation in the radial projection dimension of the target. At 0°, the radial projection dimension corresponded to the height of the cylinder, while at angle 90° the radial projection was relevant to the diameter of the cylinder.

Figure 9 shows the RCS measurement results of the metallic cylinder at 220 GHz and 440 GHz, respectively. Comparisons were made between conventional range gates and our adaptive range gates. As illustrated, for specular reflections at 0° and 180°, the difference between the two methods was negligible as the dimensions of conventional range gates and adaptive rage gates were comparable. However, notable differences were realized at 90° and 270°, corresponding to the reflection from the side face of the cylinder. The results obtained from adaptive range gates were higher. This was contributed by the reduced noise collected by the adaptive range gates, indicating improved isolation between target and clutter.

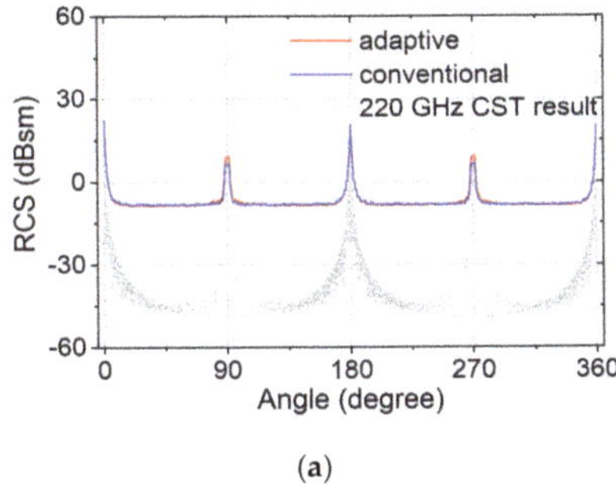

(a)

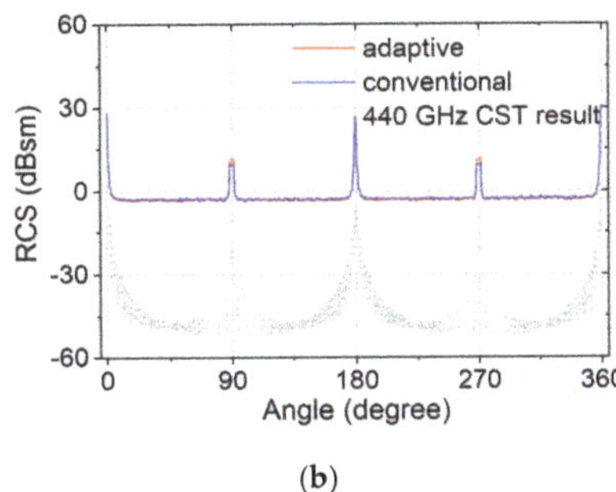

(b)

Figure 9. RCS measurement results of a metallic cylinder. (**a**) 220 GHz, (**b**) 440 GHz.

In addition, the measurement results were compared with electromagnetism-simulated RCS values, calculated by the commercial software CST, denoted by the dash lines in Figure 9. The simulation results were computed by the asymptotic solver, based on the shooting and bouncing rays (SBR) technique in the frequency domain. The solver is typically designed for efficient scattering calculation of electrically very large objects, while the contribution edge diffractions are still considered. The simulation results were first compared with theoretical RCS values of standard objects including metallic spheres, plates, cylinders, etc. Good agreements were met, confirming the reliability of the numerical calculation [21]. Compared to the calculated results, the RCS measurement errors—corresponding to the side faces—were significantly reduced by applying the adaptive range gates (2.58 dBsm at 220 GHz and 1.82 dBsm at 440 GHz), owing to the suppressed influence of the clutter. Although the absolute SNRs of the two spectra were different due to the difference in transmission power, an improvement in signal quality and measurement accuracy in both spectra were observed. This suggested an improvement in the SNR by applying the adaptive range gate method.

The effectiveness of our proposed method was further investigated by simulation. The simulation is carried out at 220 GHz. The RCS of a metallic cylinder with a large length-width ratio (height 60 cm and radius 3 cm) was calculated with the conventional range gate and adaptive range gate, respectively. The effectiveness of the adaptive range gate was quantitively evaluated by the difference in the RCS from the two methods, corresponding to the side face of the cylinder. The influences of two key factors in the adaptive range gate method were studied, *i.e.*, the dimension of the range gate and the SNR of the measurement system. Figure 10a shows the variation of the RCS difference when the size of the range gate changed from 10 points to 100 points. The 100 point range gate corresponded to the dimension of the conventional range gates, while the 10 point range gate corresponded to the minimum used in the adaptive range gate method. Its dimension was approximately 15 cm, which was sufficiently large to capture all the echoes from the target. As the dimension of the range gate increased, the RCS difference reduced. This verified that the improvement in measurement accuracy was caused by the suppressed clutter with the reduced range gate dimension. Figure 10b shows the effectiveness of our method at different SNR levels, ranging from 3 dB to 40 dB. This simulation was conducted to explore the relationship between the improvement in the RCS using the adaptive range gate method and the SNR of the system. The difference in the RCS compared to conventional range gates was slightly increased with the increasing SNR of the system, while saturated at an SNR of 25 dB. This indicated that our method could be effectively applied to RCS measurement systems regardless of its SNR.

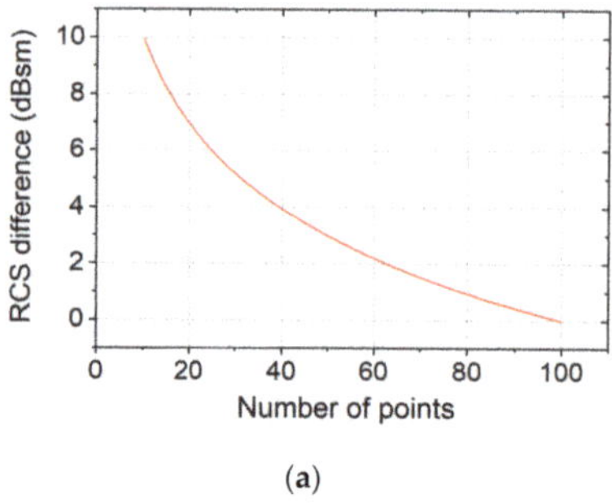

(a)

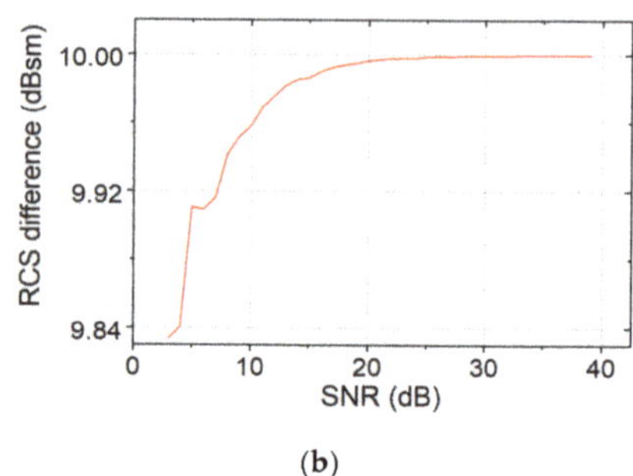

(b)

Figure 10. Comparison of RCS difference under two circumstances. (**a**) The variation of RCS difference with the size of the adaptive range gate, (**b**) the variation of RCS difference with SNR.

In the end, the adaptability of the adaptive range gate method should be discussed. In our proposed method, the dimension of the range gate was determined by the radial projection of the target. Hence, the method could be applied in RCS measurements regardless of the motion status of the target. On the other hand, in the adaptive range gate method, a linear relationship between the geometrical radial projection and the dimension of the scattering center is assumed. This is applicable for convex objects without structures causing multiple reflections. For echoes from multiple reflections, the corresponding range profile may fall out of the linear adaptive range gate, leading to insufficient echo signal capturing. Therefore, currently, the adaptability of our proposed method is limit to convex targets. Non-linear adaptive range gates should be developed for non-convex structures in the future.

5. Conclusions

In this paper, an adaptive range gate method for RCS measurement in the terahertz region was presented as a method to suppress the clutter signal and improve the SNR of RCS measurement systems. The method was developed based on a high-resolution range profile. The performance of the method was evaluated by experimental measurements, as well as numerical simulations. In our experiments, the RCS measurements of a cylinder were conducted at 220 GHz and 440 GHz, respectively. The measurement results illustrate the superiority of our method in SNR and measurement accuracy, compared to conventional fixed range gates. Simulation studies were then carried out to verify the origin of the improvement and explore the feasibility of the method. In conclusion, the adaptive range gate method was effective in improving the SNR in THz RCS measurement, especially for targets with a large length-to-width ratio. This could also work as a reference for other techniques, which require high precision scattering data of targets such as high-resolution SAR and ISAR imaging, and micro-motion detection [14]. Future development should focus on the development of non-linear adaptive range gates to extend the adaptability of the method to non-convex complex targets.

Author Contributions: S.P. conceived the adaptive range gate method, conducted the experiments and wrote the paper. Y.Z. provided theoretical guidance. Q.Y. revised the manuscript. B.D., Y.Q. and H.W. provided the experimental equipment.

Funding: This research was funded by The National Natural Science Foundation of China, grant number 61701513 and 61571011.

Conflicts of Interest: The authors declare no conflict of interest.

References

1. Knott, E.F. *Radar Cross Section*; Artech House: Dedham, MA, USA, 1985.
2. Ruan, Y.Z. *Radar Cross Section and Stealth Technology*; National Defense Industry Press: Beijing, China, 1998; pp. 6–10.
3. Knott, E.F. *Radar Cross Section Measurements*; Van Nostrand Reinhold: New York, NY, USA, 1993.

4. Dybdal, R.B. Radar Cross Section Measurements. *IEEE Trans. Antennas Propag.* **1987**, *75*, 498–516. [CrossRef]
5. Cheng, Y.Q.; Zhou, X.L.; Xu, X.W.; Qin, Y.L.; Wang, H.Q. Radar Coincidence Imaging with Stochastic Frequency Modulated Array. *IEEE J. Sel. Top. Signal Process.* **2016**, *99*, 1. [CrossRef]
6. Wang, H.Q.; Deng, B.; Qin, Y.L. Review of Terahertz Radar Technology. *J. Radars* **2018**, *7*, 1–21.
7. Chopra, N.; Yang, K.; Abbasi, Q.H.; Qaraqe, K.; Philpott, M.; Alomainy, A. THz Time Domain Spectroscopy of Human Skin Tissue for In-Body Nano-networks. *IEEE Trans. Terahertz Sci. Technol.* **2016**, *6*, 803–809. [CrossRef]
8. Iwaszczuk, K.; Heiselberg, H.; Jepsen, P. Terahertz Radar Cross Section Measurements. *Opt. Express* **2010**, *18*, 26399–26408. [CrossRef] [PubMed]
9. Coulombe, M.J.; Horgan, T.; Waldman, J.; Szatowski, G.; Nixon, W. A 524 GHz Polarimetric Compact Range for Scale Model RCS Measurements. In Proceedings of the Antenna Measurement Techniques Association 21st Annual Meeting & Symposium (AMTA '99), Monterey Bay, CA, USA, 4–8 October 1999; pp. 458–463.
10. Samoska, L.A. An Overview of Solid-State Integrated Circuit Amplifiers in the Submillimeter-Wave and THz Regime. *IEEE Trans. Terahertz Sci. Technol.* **2011**, *1*, 9–24. [CrossRef]
11. Beaudoin, C.J.; Horgan, T.; Demartinis, G.; Coulombe, M.J.; Goyette, T.; Gatesman, A.J.; Nixon, W.E. A Prototype Fully Polarimetric 160-GHz Bistatic ISAR Compact Radar Range. In Proceedings of the SPIE 10188, Radar Sensor Technology XXI, Anaheim, California, CA, USA, 1 May 2017.
12. Cooper, K.B.; Dengler, R.J.; Chattopadhyay, G.; Schlecht, E.; Gill, J.; Skalare, A.; Mehdi, I.; Siegel, P.H. A High-Resolution Imaging Radar at 580 GHz. *IEEE Microw. Wirel. Compon. Lett.* **2008**, *18*, 64–66. [CrossRef]
13. Macfarlane, D.G.; Robertson, D.A.; Bryllert, T. Pathfinder—A High Resolution 220 GHz Imaging Radar Providing Phenomenological Data for Security Scanner Development. In Proceedings of the 2016 41st International Conference on Infrared, Millimeter, and Terahertz Waves (IRMMW-THz), Copenhagen, Denmark, 25–30 September 2016.
14. Yang, Q.; Qin, Y.L.; Deng, B.; Wang, H.Q.; You, P. Micro-Doppler Ambiguity Resolution for Wideband Terahertz Radar Using Intra-Pulse Interference. *Sensors* **2017**, *17*, 993. [CrossRef] [PubMed]
15. Chen, G.; Dang, H.X.; Tan, X.M.; Chen, H.; Cui, T.J. Scattering Properties of Electromagnetic Waves from Randomly Oriented Rough Metal Plate in the Lower Terahertz Region. *J. Radars* **2018**, *7*, 75–82.
16. Song, Y.; Li, Y.P.; Pang, S.; Zhao, S.S.; Wang, H.Q. RCS Measurement at Terahertz Waves for Cylinders with Different Surface Roughness. *Electron. Lett.* **2018**, *54*, 714–716. [CrossRef]
17. Choe, W.; Jeong, J. A Broadband THz On-Chip Transition Using a Dipole Antenna with Integrated Balun. *Electronics* **2018**, *7*, 236. [CrossRef]
18. Siegel, P.H.; de Maagt, P.; Zaghloul, A.I. Antennas for Terahertz Applications. In Proceedings of the 2006 IEEE Antennas and Propagation Society International Symposium, Albuquerque, NM, USA, 9–14 July 2006.
19. Zhang, X.L.; Li, N.J.; Hu, C.F.; Li, P. *Scattering Characteristics Test and Imaging Diagnosis for Targets in Radar System*; China Aerospace Publishing House: Beijing, China, 2009; pp. 1–47.
20. Hu, C.F. Research on RCS Measurement System and Microwave Imaging Diagnosis Technology. Master's Thesis, Northwestern Polytechnical University, Xi'an, China, March 2007.
21. Li, Z.; Cui, T.J.; Zhong, X.J.; Tao, Y.B.; Lin, H. Electromagnetic Scattering Characteristics of PEC Targets in the Terahertz Regime. *IEEE Antennas Propag. Mag.* **2009**, *51*, 39–50.

Article

Accurately Modeling of Zero Biased Schottky-Diodes at Millimeter-Wave Frequencies

Jéssica Gutiérrez [1], Kaoutar Zeljami [2], Tomás Fernández [3,*], Juan Pablo Pascual [3] and Antonio Tazón [3]

1 Erzia Technologies, C/Josefina de la Maza 4 - 2°, 39012 Santander, Spain; jessica.gutierrez@erzia.com
2 Information and Telecommunication Systems Laboratory, Université Abdelmalek Essaâdi, Tetouan 2117, Morocco; kaoutar_ele@hotmail.com
3 Department of Communications Engineering, Laboratorios de Ingenieria de Telecomunicación Profesor José Luis García García, University of Cantabria, Plaza de la Ciencia, 39005 S/N Santander, Spain; juanpablo.pascual@unican.es (J.P.P.); antonio.tazon@unican.es (A.T.)
* Correspondence: tomas.fernandez@unican.es; Tel.: +34-942-200-887

Received: 14 May 2019; Accepted: 17 June 2019; Published: 20 June 2019

Abstract: This paper presents and discusses the careful modeling of a Zero Biased Diode, including low-frequency noise sources, providing a global model compatible with both wire bonding and flip-chip attachment techniques. The model is intended to cover from DC up to W-band behavior, and is based on DC, capacitance versus voltage, as well as scattering and power sweep harmonics measurements. Intensive use of 3D EM (ElectroMagnetic) simulation tools, such as HFSS™, was done to support Zero Biased Diode parasitics modeling and microstrip board modeling. Measurements are compared with simulations and discussed. The models will provide useful support for detector designs in the W-band.

Keywords: W band; Schottky Diode Detectors; ZBD modeling; wire bonding; flip-chip

1. Introduction

Schottky diodes play an important role in several functions, such as rectification [1,2], mixing [3,4], and detection in all the range of microwave frequencies, up to the W-band and beyond [4].

In general, modeling of the diodes is a critical task in the design process of circuits operating at high frequencies [3,4]. Antimony (Sb) heterostructure backward diodes offer better noise performance with easier matching to 50 Ohm [5] compared to GaAs Schottky diodes, but may not always be easily accessible or available as discrete components [6]. Series 9161 diodes (Keysight HSCH and MACOM-Metelics MZBD) models are frequently used for hybrid detectors below the W-band [7–9], though also in the W-band [10–13]. Zero-bias Schottky diodes from ACST (Advanced Compound Semiconductor Technologies GmbH) (https://acst.de/) operating as power sensors have been reported in [14]. Zero Bias Diodes Schottky diodes from Virginia Diode Inc. (VDI) were used for W-band detection in [6,15]. Available manufacturer information (https://www.vadiodes.com/images/pdfs/Spec_Sheet_for_VDI_W_Band_ZBD.pdf) is limited and, in the absence of diode models developed by the designers themselves, commercially available models could be required.

Usually, the device modeling is accomplished by discriminating between the modeling of the intrinsic component (characterized by quasi-static I-V and C-V measurements) and the extrinsic elements (characterized by 3D Electro-Magnetic (EM) tools, along with the measurement of the Scattering parameters at some specific bias points). In [16], modeling techniques are presented to extract a Schottky diode model analytically using additional fabricated structures (such as short and open circuits) for the de-embedding of parasitic capacitances and other effects. Unfortunately, it is not always possible to have such customized cal kit (calibration kit) structures available. Previous efforts in Schottky diode modeling of Single Anode (SA) VDI devices were presented in [17]. In this paper, ZBD (Zero Bias Diode) devices from VDI have been carefully modeled, considering two different diode-mounting techniques to attach them to the final circuit assembly: wire bonding and flip-chip (also known as controlled collapse chip connection). From DC measurements, the intrinsic nonlinear current source, along with the parasitic resistor can be obtained. Scattering measurements of the ZBD provide data which can be used to identify the parasitic and extrinsic elements up to the W-band; in this sense, in Section 2, the modeling of ZBDs for wire bonding assembling is presented. In Section 3, the modified model oriented to the flip-chip attachment technique, obtained from the previously extracted diode model, is shown; low-frequency noise sources are included in the complete diode model in Section 4. Finally, some conclusions are drawn in Section 5.

2. Nonlinear Zero Bias Diode Modelling

Modeling of the ZBD was based on the procedures presented in [16,17]. Firstly, a 3D model of the passive parts for EM simulation (HFSS^{TM}) was built based on microphotographs (Figure 1) and physical measurements of the critical dimensions, performed with a Scanning Electron Microscope (SEM) available in the Laboratory of Science and Engineering of Materials of the University of Cantabria (LADICIM).

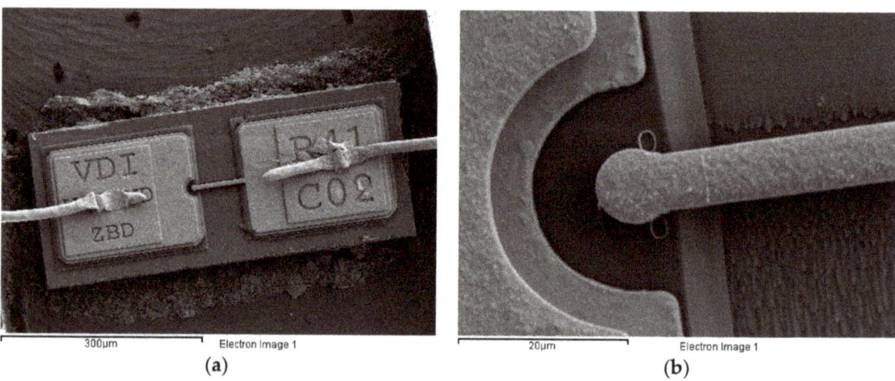

(a) (b)

Figure 1. (**a**) General View of the ZBD using SEM (Dice about 600 × 250 μm; anode finger length about 100 μm). (**b**) Detail of the anode finger of the ZBD (diameter about 9 μm). Reference scale is at the bottom of the figures.

2.1. DC Measurements and Nonlinear Model

For modeling purposes, as a starting point, the diode equivalent circuit presented in Figure 2 is here considered. In this circuit, the main non-linearities (Id Current Source and Cj Capacitance), as well as parasitic elements due to metallization (Rs) and packaging (Cpp, Cp, and Lf), can be observed.

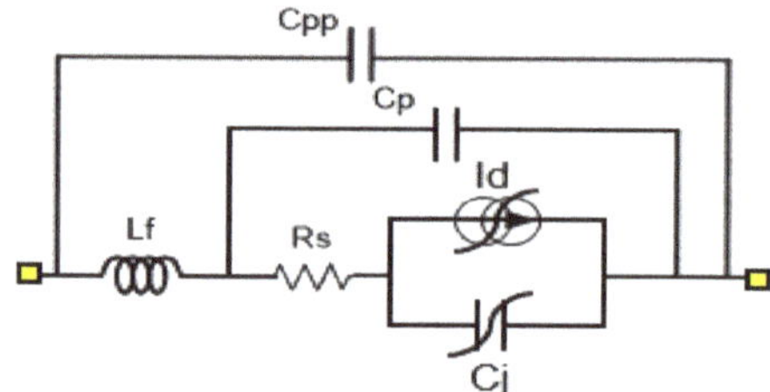

Figure 2. Diode equivalent circuit model.

Using a specific high-accuracy voltage-controlled current source to ensure both protection of the device during the test process and accurate current/voltage measurements in the low current level region (below 1 mA), the static I/V characteristic of the ZBD was measured (Figure 3). By means of proprietary extraction software, the value of the thermionic-field emission model parameters [18] for the nonlinear current source (1) were obtained (Table 1).

$$I(V_d, T) = I_s e^{\left(\frac{q(V_d - I_d R_S)}{\eta K T}\right)} \tag{1}$$

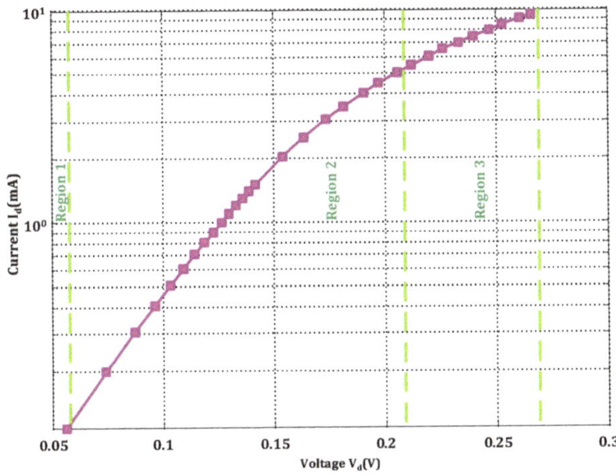

Figure 3. I-V measured current for the ZBD Schottky diode under testing.

Table 1. Extracted value of parameters in Equation (1).

Parameter	Value
I_s (A)	2.78×10^{-5}
η	1.37
$\alpha = \frac{q}{\eta k T}$ (1/V)	28.6
R_s (Ω)	6.48

2.2. *Low-Frequency Model*

At lower microwave frequencies, the equivalent circuit of the unbiased, or negatively biased, diode can be reduced to a "π" electrical network of capacitances, as depicted in Figure 4.

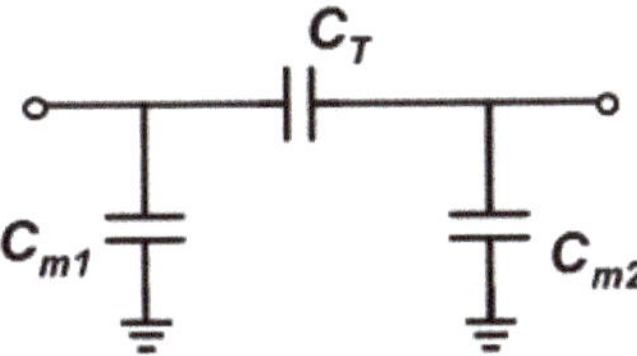

Figure 4. Low-frequency equivalent circuit model of the passive diode.

The above "π" network accounts for the total capacitance C_T, as well as the small-value parasitic capacitances to the ground (if considered), C_{m1} and C_{m2}, where C_T is the sum of the junction capacitance C_j and the parasitic capacitances C_{pp}, as shown in Equation (2).

$$C_T = C_j(V) + C_{pp} = \frac{C_{j0}}{\sqrt{1 - \frac{V_d}{\phi_{bi}}}} + C_{pp} \tag{2}$$

An accurate estimation of the total capacitance can be extracted from the measured value of the S_{21} scattering parameter (Figure 5) when a negative bias, V_d, is applied to the diode. Parameter S_{21} essentially depends on the total capacitance, C_T (the junction capacitance C_j at V_d = 0 V, C_{j0}, junction potential, ϕ_{bi}, and the parasitic capacitance C_{pp}). Scattering parameters of the diode were measured from 2 GHz up to 50 GHz; however, for capacitance extraction purposes, only the frequency range of 3–10 GHz was used, considering the equivalent circuit in Figure 4. By varying the negative V_d voltage applied to the diode, and performing the measurement of the scattering parameters of the diode at each voltage, it was possible to extract the values of the C_T capacitance as a function of this voltage. In Figure 6, the extracted Capacitance versus Voltage (*C-V*) plot is presented. In Table 2, the values of the elements of the circuit in Figure 2, as well as the parameters of Equation (2), are presented.

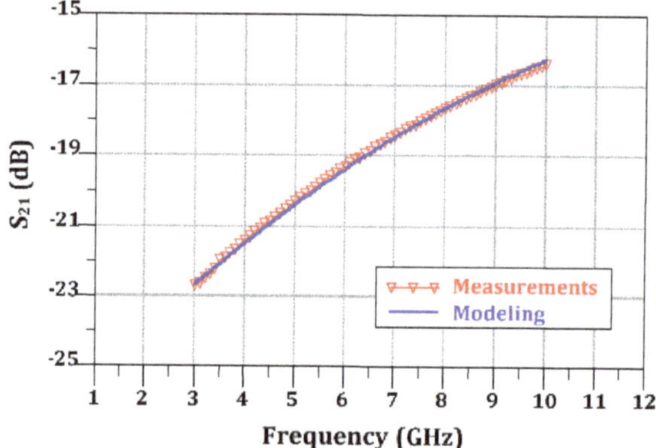

Figure 5. Measurements and modelling of transmission coefficient for the unbiased (V_d = 0 V) ZBD.

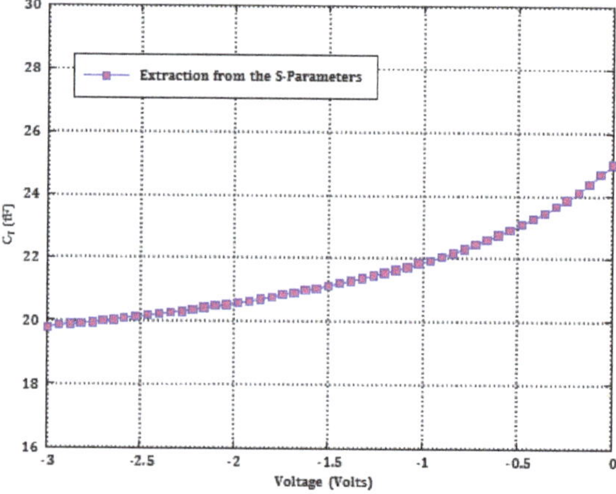

Figure 6. Results for the extraction of total capacitance of the ZBD from scattering parameters.

In principle, the results obtained in the extraction from the S-parameters offer more accurate estimations of nonlinear and parasitic capacitances of this diode, compared to other extraction methods based on low-frequency impedance measurements. This also avoids other potential problems of low-frequency measurements, such as uncertainty in the de-embedding of the parasitic effects of probes.

Table 2. Value of parameters of Equation (2).

Parameter	Value
C_T (fF)	26
C_{j0} (fF)	10
C_{pp} (fF)	16
ϕ_{bi} (V)	0.16
C_{m1} (fF)	50
C_{m2} (fF)	50

2.3. Modified ZBD Model Up to W-Band

At high frequencies, from 75 GHz up to 110 GHz, additional parasitic effects arise which affect the behavior of the diode; thus, it is critical to evaluate and model them. This is really a complex task due to the difficulties involved in the de-embedding process of the access elements, such as coplanar-to-microstrip (CPW-M) transitions and bonding wires used in the circuit assembly to implement the input and output connections. This is the reason why, although the diode is usually considered as a one-port element, with the cathode or the anode grounded, we are going to perform measurements being anode and cathode the electric ports, considering the diode as a two-port electrical network, allowing to identify the effect of grounding.

Firstly, modeling of the coplanar to microstrip (CPW-M) transitions was performed, as reported in [17]. Once the model of the CPW-M transition was obtained, the modeling of the ZBD, when using the wire bonding attachment technique, was possible from the measurements performed at the coplanar probe station using a LRRM (Line-Reflect-Reflect-Match) calibration. A de-embedding procedure in Keysight ADS™ [19] was performed considering the previously obtained model of the CPW-M transition along with the gold bonding wire model.

In Figure 7, the reflection and transmission coefficients from the equivalent circuit simulation are compared with the experimental measurements in two frequency ranges (2–50 GHz and 75–110 GHz) for the unbiased diode (V_d = 0 V). From these results, it can be concluded that the degree of agreement between both results is excellent. Therefore, the modeling procedure proposed provides a model that can simulate the behavior of ZBD from the DC up to 110 GHz.

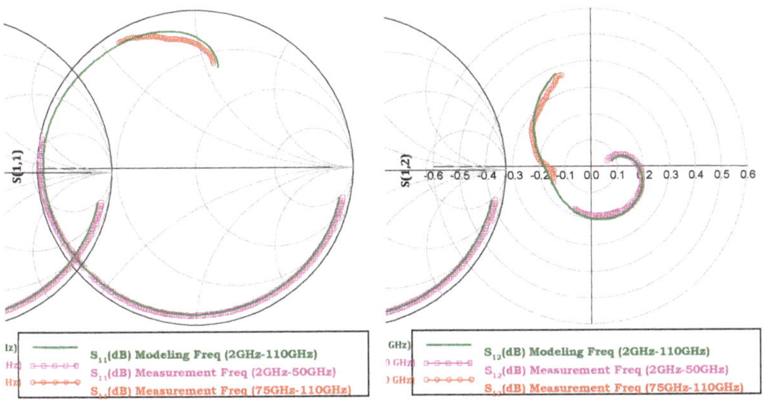

Figure 7. *Cont.*

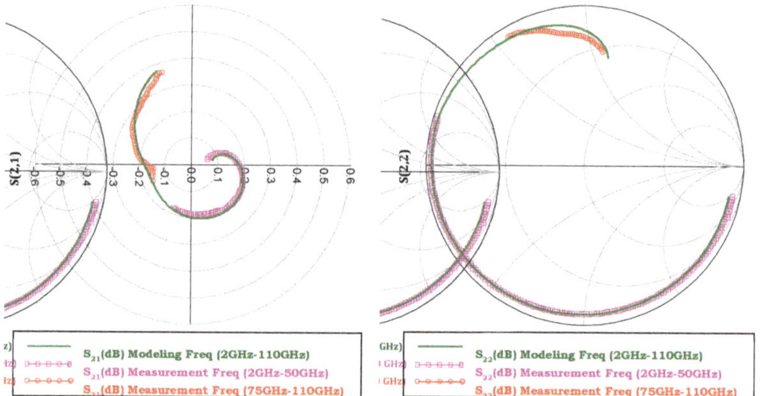

Figure 7. Modelled and measured [S]-parameters of the unbiased diode (0 V) after the de-embedding in the wire bonding configuration.

An increase in the series resistance value was observed in the W-band measurements; this singular behavior can be associated with the skin effect above 50 GHz, the bonding wires, and probably some other measurement uncertainties. The study of the extent of this effect becomes difficult in this frequency band as measurements are limited by calibration accuracy, as well as some possible inaccuracies in the determination of other equivalent circuit parameters, such as finger inductance and parasitic capacitances. Nevertheless, the increase in resistance in the W-band was consistently found when fitting measurements of several devices, as previously reported in the literature [20].

2.4. Large Signal Model Extraction and Validation

The final topology of the proposed diode's large signal model is shown in Figure 8 [6].

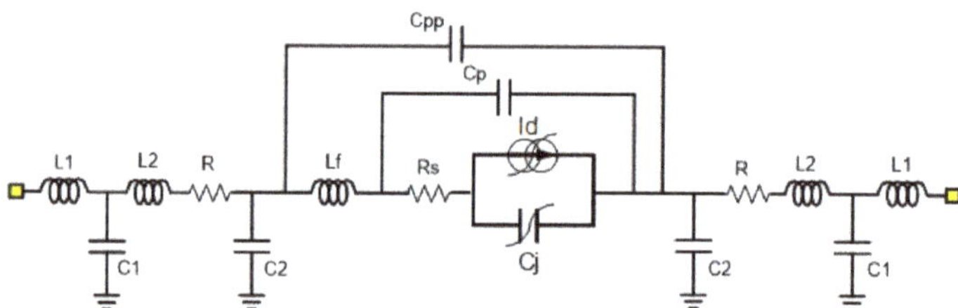

Figure 8. Equivalent circuit model of the diode, including the extrinsic elements.

In Table 3, the values of the model parameters obtained from the different measurements performed using the extraction procedures [16,17] are presented. In this model, C_p represents the finger-to-pad capacitance between the anode contact finger and the underlying active GaAs. The bonding resistance, R is fixed to 0 Ohm at "low frequencies", while its value equals to 3 Ohm in the W-band.

In order to validate the large-signal model of the diode, output power versus input power measurements were compared with harmonic balance simulations performed in Keysight ADS™ [19]. In Figure 9, comparisons between measurements and simulations, when a signal of 1 GHz is applied to the device, are depicted. In this graph, a reasonable global agreement between measurements and simulation results can be seen at the fundamental frequency and the second and third order harmonics. If we focus on the highest input power level, simulations tend to overestimate the first harmonic, while the second is more accurate as the power is higher. The third harmonic is well-estimated in values, but with slight differences in the slope. This could be due to several reasons: the source current mode was originally intended to simulate the nonlinear current source in DC conditions;

however, the large-signal behavior depends on the model's ability to predict harmonics, something that, in general, is difficult to guarantee as it requires a more elaborate development of the nonlinear current source model. Furthermore, the influence of nonlinear capacitance should be considered, which could require a more accurate characterization at the highest power levels.

Table 3. Extracted values of the parameters for the ZBD model.

Parameter	Symbol	Value	Parameter	Symbol	Value
Saturation Current (A)	I_{sat}	28×10^{-6}	Parasitic Finger Capacitance (fF)	C_p	1
Ideality Factor	η	1.4	Pad-to-Pad Capacitance (fF)	C_{pp}	16
$\alpha = \frac{q}{\eta kT}$ (1/V)	α	28.6	Bonding Inductance (pH)	L_1	73
Junction Capacitance ($V_d = 0$ V) (fF)	C_{j0}	10	Bonding Inductance (pH)	L_2	42
Series Resistance (Ω)	R_s	6.5	Pad Capacitance (fF)	C_1	88
Junction Potential (V)	ϕ_{bi}	0.16	Pad Capacitance (fF)	C_2	51
Finger Inductance (pH)	L_f	50	Bonding Resistance (Ω)	R	0 (Low Frequency) 3 (W Band)

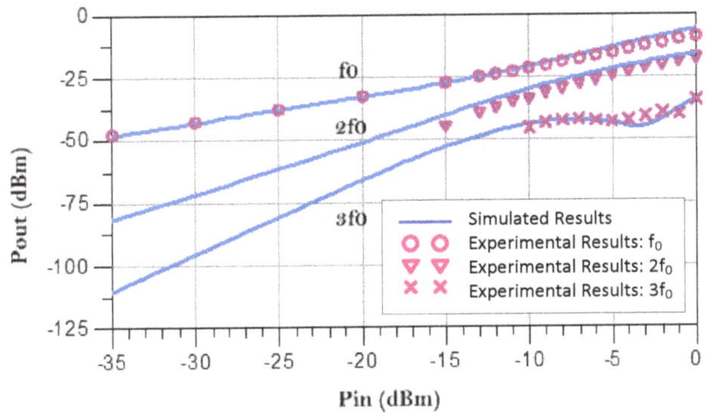

Figure 9. Simulated (continuous line) and measured output power versus input power. (Circles: f_0, Inverted Triangle: $2f_0$, Cross: $3f_0$).

3. Diode Modelling Oriented to Flip-Chip Attachment Technique

The previous model is a basic general-purpose model, but particularly considering W-band applications, it was necessary to achieve a better approximation for the most commonly used attachment technique of the ZBD to the circuit assembly, while being well-based on the previously presented model extraction. Therefore, a new circuit was assembled by attaching a ZBD using the flip-chip technique, as shown in Figure 10. This attachment technique is the most critical in terms of the location of the diode, in order to achieve a symmetrical structure and to avoid destroying the finger of the diode by placing it in contact with the CPW-M transition. Please note that, to attach the flip-chip into a circuit, the chip must be inverted to bring the solder dots (made of H20E silver epoxy) down onto the connectors on the circuit board; then, the solder is re-melted (120 °C for 15 min) to produce an electrical connection, leaving a small space between the diode and the underlying mounting.

Figure 10. Photo of the ZBD in the flip-chip assembly.

As performed for the previously presented modeling procedure, CPW-M transitions were used, and their model was used in the optimization in Keysight ADS^{TM}, in order to obtain an equivalent circuit model consisting only of the intrinsic and parasitic elements of the diode. A comparison between experimental measurements and the model (described below) simulation results is shown in Figure 11, showing a better match at the lower frequencies of the W-band.

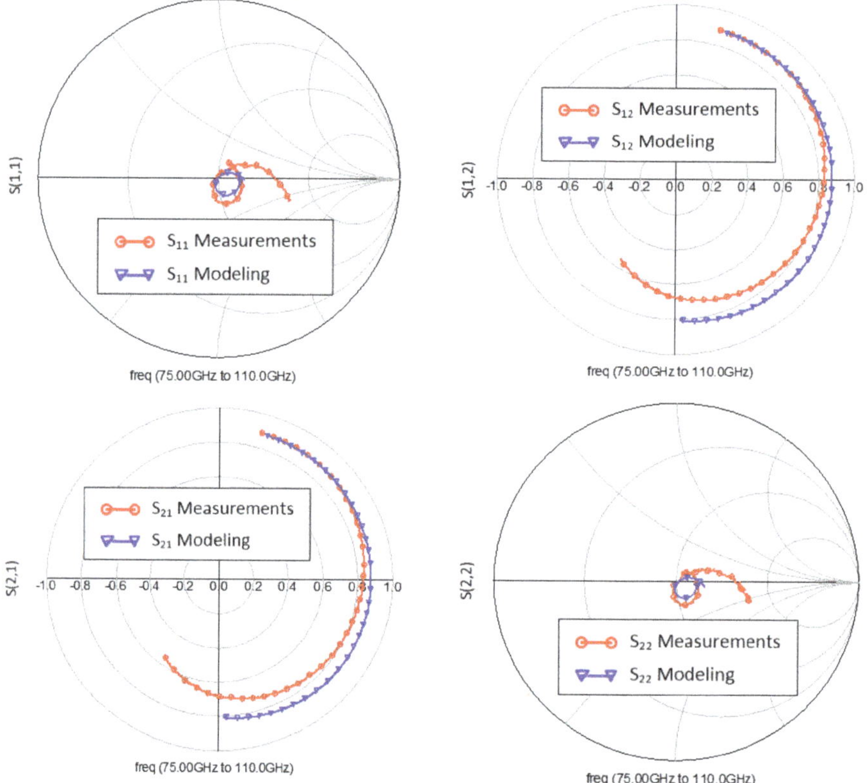

Figure 11. Comparison between S-parameter measurements and modeling simulation of the unbiased ZBD (0 V) for the flip-chip assembly in the W-band.

Flip-Chip Zero Bias Diode Complete Model

The resulting equivalent circuit model is presented in Figure 12. Comparing this equivalent circuit with the previous model (Figure 8), the main difference is that some capacitances have been added to account for some newly existing parasitic effects. In the flip-chip assembly, the diode was attached to the carrier upside down, compared to the conventional mode with bonding wires. Therefore, capacitances C_2 in Figure 8 are identified in Figure 12 as C_{PAD1} and C_{PAD2}, being connected to a new capacitance, C_{GNDs}, which represents a new, slightly capacitive coupling effect between the microstrip reference plane of the diode bulk and the ground plane of the total circuit assembly. An additional capacitive coupling effect is thus originated, $C_{COUPLING}$, which is related to the capacitive coupling between the CPW-M transitions since, in this assembly, both transitions are placed closer than in the previous one. The L_{IN}-C_{IN}-R_{IN} and R_{OUT}-C_{OUT}-L_{OUT} electrical networks simulate the microstrip line effect at the connection between the diode and the CPW-M transition; besides, R_{IN} and R_{OUT} also include the losses related to the H20E conductive silver epoxy used to attach the diode. It should be emphasized that fitting in Figure 7 covers the range from 2 to 110 GHz, de-embedding the effect of the wire bonding, while Figure 11 covers a frequency range from 75 to 110 GHz, comparing measurements and simulations without de-embedding. The upside-down positioning of the diode causes some uncertainty in the parasitic capacitances with the ground, but at the same time, other parasitic elements must be maintained for coherence with grounded-microstrip ZBD model (Figure 8). Those restrictions limit the capability for a perfect impedance matching.

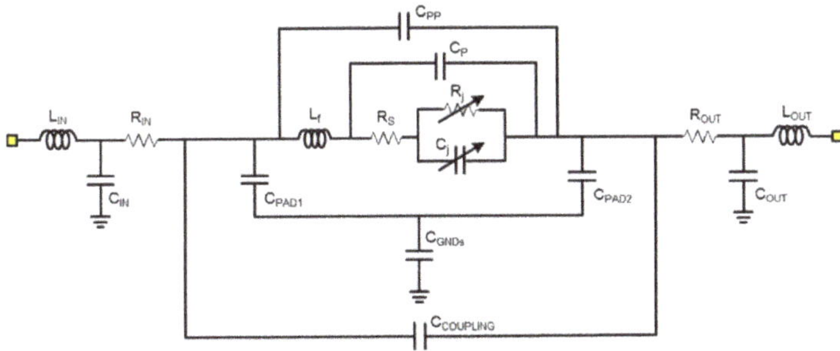

Figure 12. Equivalent circuit model for the ZBD in flip-chip assembly.

The resulting model parameters extracted for this second diode assembly technique are listed in Table 4.

Table 4. Results of the parameter extraction of the total capacitance of ZBD.

Parameter	Symbol	Value
Saturation Current (A)	I_{sat}	28×10^{-6}
Ideality Factor	η	1.4
$q/\eta kT$ (1/V)	α	28.6
Junction Capacitance (V_d = 0 V) (fF)	C_{j0}	10
Series Resistance (Ω)	R_S	6.5
Junction Potential (V)	ϕ_{bi}	0.16
Finger Inductance (pH)	L_f	50
Parasitic Capacitance (fF)	C_P	1
Pad-to-Pad Capacitance (fF)	C_{PP}	16
Input Inductance (pH)	L_{IN}	59
Output Inductance (pH)	L_{OUT}	48
Input Capacitance (fF)	C_{IN}	3.4
Output Capacitance (fF)	C_{OUT}	0.3
Input Resistance (Ω)	R_{IN}	2.3
Output Resistance (Ω)	R_{OUT}	2.3
Coupling Capacitance (fF)	$C_{COUPLING}$	18
PAD IN Capacitance (fF)	C_{PAD1}	39
PAD OUT Capacitance (fF)	C_{PAD2}	203
Ground Capacitance (fF)	C_{GNDs}	11

4. Including Low-Frequency Noise in the Zero-Bias Model

For a complete evaluation of diode performance, which is of significant use in detectors, a knowledge of low-frequency noise behavior becomes quite relevant—not only as proof of technology quality, but also to establish the noise floor, which limits the dynamic range. Thus, this ensures a stable response in a dynamic mode of operation. In this sense, a set-up was developed to measure and model low-frequency noise sources present in the ZBD, following [9]. The set-up diagram is shown in Figure 13, and the corresponding photo is shown in Figure 14. In this case, the diode was mounted in the conventional way (grounded, not Flip-Chip) inside a shielded box to avoid unwanted interference in the measurements. In this low-frequency range, a distinction between the flip-chip and wire bonding is not relevant.

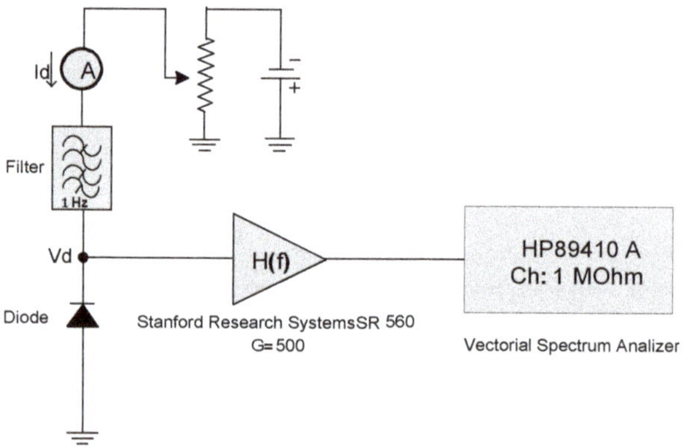

Figure 13. Setup to evaluate low-frequency noise under different DC biases.

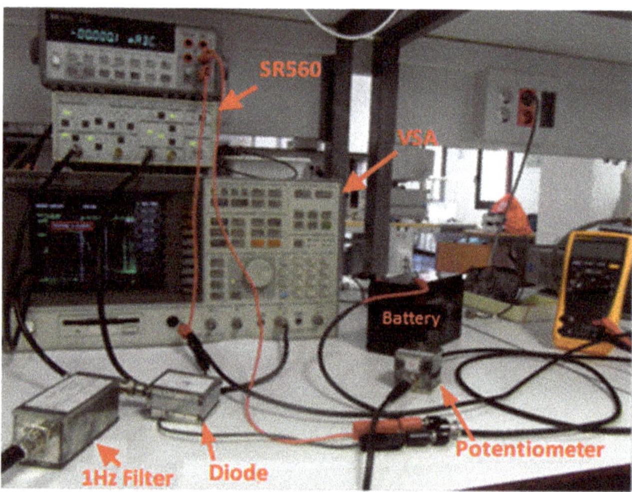

Figure 14. Photo of the measurement setup for low-frequency noise, corresponding to the diagram in Figure 13.

The main low-frequency noise contributions expected in the current spectral density (CSD) were: thermal noise associated with the series resistance R_s, shot noise, and flicker noise. The equations governing the different noise sources are given in Equations (3)–(5) (where K is the Boltzmann constant, f is the low frequency, q is electron charge, I_s is the saturation current, and I_{diode} is the actual current, whereas K_f, a_f, and b_f are parameters to be determined). Noise sources were placed in the circuital model of the diode, as shown in Figure 15.

$$\langle CSD_{thermal} \rangle = \overline{i_{nR_s}^2} = \frac{4KT}{R_s} \tag{3}$$

$$\langle CSD_{shot} \rangle = \overline{i_{nshot}^2} = 2q(I_{diode} + 2I_s) \tag{4}$$

$$\langle CSD_{flicker} \rangle = \overline{i_{n_{1/f}}^2} = k_f \frac{I^{a_f}}{f^{b_f}} \tag{5}$$

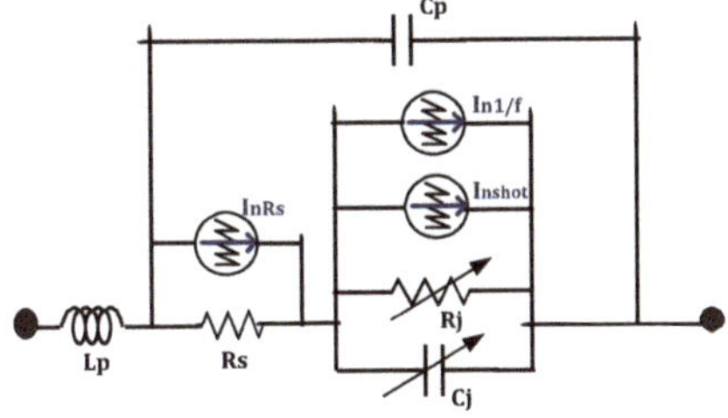

Figure 15. Insertion of noise sources in the diode model.

The low-frequency noise spectrum measured with the set-up in Figures 13 and 14 is shown in Figure 16, including the system noise floor (NF trace) and measurements for the four different current bias points: 0.05 mA, 0.1 mA, 0.15 mA, and 0.2 mA. The set-up was also simulated considering all the noise contributions and Flicker noise parameters in Equation (5), which were optimized to fit with the measurements at these four bias currents. In Table 5, the optimized parameters are listed.

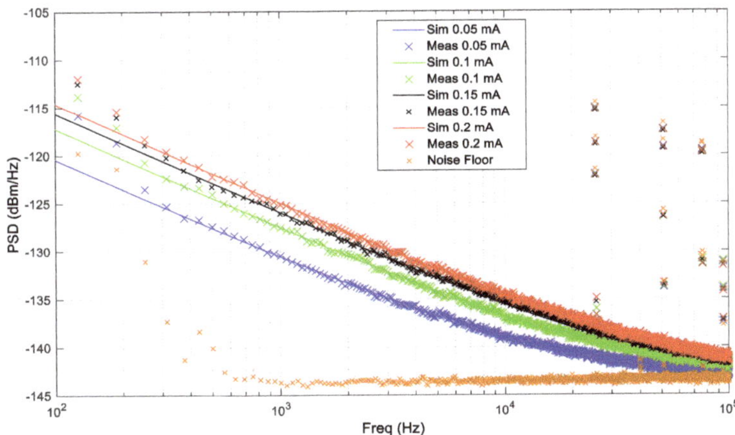

Figure 16. Low-frequency noise spectrum measurements of the set-up, shown in Figure 13, and simulations using the model in Figure 15.

Table 5. Low-frequency noise parameters.

Parameter	Value
k_f	1.08×10^{-7}
a_f	2.48
b_f	1.04

5. Conclusions

In this paper, two models for Zero-Biased Diodes (VDI) considering wire-bonded or flip-chip techniques have been presented, with particular emphasis on the flip-chip model in the W-band. Equivalent circuit models were complemented with the extraction of low-frequency noise sources (Flicker) and its incorporation in the proposed models, allowing for a more complete prediction of the response of the diodes, for example, in detectors. The obtained models were validated under different operating conditions, and power prediction abilities of the model were probed performing nonlinear harmonic measurements ranging the appropriate power levels, in comparison with harmonic balance simulations in commercial software (Keysight ADS™).

Comparisons up to the W-band between measured and simulated scattering parameters using the complete model and showing a good agreement were performed, probing the validity of the extraction methods applied using both wire bonding and flip-chip attachment techniques.

Finally, low-frequency noise measurements and simulations were performed to validate the noise model added to the previously extracted diode model, which showed an adequate level of agreement.

As a general remark in "normal" operations, Zero Bias diodes are expected to be self-biased by the input RF power, as it happens in the measurements shown in Figure 9, where input power was swept. In Figure 3, ZBD diodes were biased for DC fitting, and scattering parameters were also measured in low microwave frequencies under a DC bias to fit the capacitance model (Figures 5 and 6). For low-frequency noise measurements also, the DC bias was applied to put into the evidence bias dependence of the Flicker noise parameters. We considered that those measurements would cover the usual scenarios for ZBD operations. A combined application of W-band power and DC bias to the Zero Bias diodes was avoided, keeping in mind the fragility of the diode because of its low-threshold voltage, where voltages applied any higher than the safe one would totally or partially damage it, leading to unreal measurements of the device's high-frequency behavior.

Author Contributions: Conceptualization, J.P.P. and T.F.; methodology, J.G. and K.Z.; software, J.G.; validation, J.G. and K.Z.; formal analysis, J.G. and T.F.; investigation, J.G. and K.Z.; writing—original draft preparation, J.P.P. and T.F.; writing—review and editing, J.P.P. and T.F.; supervision, A.T.; project administration, J.P.P.; funding acquisition, A.T.

Funding: This research was funded by the Spanish Ministry of Economy, Science and Innovation for the financial support provided through projects CONSOLIDER-INGENIO CSD2008-00068 (TERASENSE), the continuing excellence network SPATEK and the projects TEC2014-58341-C4-1-R and TEC2017-83343-C4-1-R.

Acknowledgments: The authors would like to thank the University of Cantabria Industrial Doctorate program 2014, project: "Estudio y Desarrollo de Tecnologías para Sistemas de Telecomunicación a Frecuencias Milimétricas y de Terahercios con Aplicación a Sistemas de Imaging en la Banda 90–100 GHz". The authors would like also to express their gratitude to all the staff of DICOM's Microwaves & RF group for their help and to the assembling laboratory staff: Ana Pérez, Eva Cuerno and Sandra Pana for their help with the fabrication of the prototypes and to Dermot Erskine for the revision of the text.

Conflicts of Interest: The authors declare no conflict of interest. The funders had no role in the design of the study; in the collection, analyses, or interpretation of data; in the writing of the manuscript, or in the decision to publish the results.

References

1. Biebl, E.M. RF Systems Based on Active Integrated Antennas. *Int. J. Electron. Commun.* **2003**, *57*, 173–180. [CrossRef]
2. Shi, Y.; Jing, J.; Fan, Y.; Yang, L.; Li, Y.; Wang, M. A novel compact broadband rectenna for ambient RF energy harvesting. *Int. J. Electron. Commun.* **2018**, *95*, 264–270. [CrossRef]
3. Janin, S.; Sripimanwat, K.; Phongcharoenpanich, C.; Krairiksh, M. A hybrid ring coupler quasi-optical antenna-mixer. *Int. J. Electron. Commun.* **2009**, *63*, 36–45. [CrossRef]
4. Yao, C.; Chen, Z.; Ge, J.; Zhou, M.; Wei, X. A compact 220 GHz heterodyne receiver module with planar Schottky diodes. *Int. J. Electron. Commun.* **2018**, *84*, 153–161. [CrossRef]

5. Moyer, H.P.; Schulman, J.N.; Lynch, J.J.; Schaffner, J.H.; Sokolich, M.; Royter, Y.; Bowen, R.L.; McGuire, C.F.; Hu, M.; Schmitz, A. W-Band Sb-Diode Detector MMICs for Passive Millimeter Wave Imaging. *IEEE Microw. Wirel. Compon. Lett.* **2008**, *18*, 686–688. [CrossRef]
6. Hrobak, K.M.; Sterns, M.; Schramm, M.; Stein, W.; Schmidt, L.P. Planar zero bias Schottky diode detector operating in the E- and W-band. In Proceedings of the International 2013 European Microwave Conference, Nuremberg, Germany, 6–10 October 2013; pp. 179–182. [CrossRef]
7. Artal, E.; Aja, B.; de la Fuente, M.L.; Pascual, J.P.; Mediavilla, A.; Martinez-Gonzalez, E.; Pradell, L.; de Paco, P.; Bara, M.; Blanco, E.; et al. LFI 30 and 44 GHz receivers Back-end Modules. *J. Instrum.* **2009**, *4*, T12003. [CrossRef]
8. Cano, J.L.; Aja, B.; Villa, E.; de la Fuente, L.; Artal, E. Broadband back-end module for radio-astronomy applications in the Ka-Band. In Proceedings of the 38th European Microwave Conference, Amsterdam, The Netherlands, 27–31 October 2008; pp. 1113–1116, ISBN 978-2-87487-006-4. [CrossRef]
9. Gutiérrez, J.; Zeljami, K.; Villa, E.; Aja, B.; de la Fuente, M.L.; Sancho, S.; Pascual, J.P. Noise conversion of Schottky diodes in mm-wave detectors under different nonlinear regimes: Modeling and simulation versus measurement. *Int. J. Microw. Wirel. Technol.* **2016**, *8*, 479–493. [CrossRef]
10. Xie, L.; Zhang, Y.; Fan, Y.; Xu, C.; Jiao, Y. A W-band Detector with High Tangential Signal Sensitivity and Voltage Sensitivity. In Proceedings of the 2010 International Conference on Microwave and Millimeter Wave Technology (ICMMT), Chengdu, China, 8–11 May 2010; pp. 528–531.
11. Xu, K.; Zhang, Y.; Xie, L.; Fan, Y. A Broad W-band Detector Utilizing Zero-bias Direct Detection Circuitry. In Proceedings of the 2011 International Conference on Computational Problem-Solving (ICCP), Chengdu, China, 21–23 October 2011; pp. 190–194.
12. Yao, C.; Zhou, M.; Luo, Y.; Xu, C. Millimeter wave broadband high sensitivity detectors with zero-bias Schottky diodes. *J. Semicond.* **2015**, *36*. [CrossRef]
13. Tekbaş, M.; Erdoğan, M.S.; Ünal, İ. A W band waveguide detector module using zero bias schottky diode. In Proceedings of the 2017 IEEE 37th International Conference on Electronics and Nanotechnology (ELNANO), Kiev, Ukraine, 18–20 April 2017; pp. 137–142. [CrossRef]
14. Hoefle, M.; Penirschke, A.; Cojocari, O.; Jakoby, R. Broadband zero-bias Schottky detector for E-field measurements up to 100 GHz and beyond. In Proceedings of the 2013 38th International Conference on Infrared, Millimeter, and Terahertz Waves (IRMMW-THz), Mainz, Germany, 1–6 September 2013; pp. 1–2. [CrossRef]
15. Zhang, W.; Yang, F.; Wang, Z.X. W-band (90GHz) zero bias Schottky diode directive detector. In Proceedings of the 2015 Asia-Pacific Microwave Conference (APMC), Nanjing, China, 6–9 December 2015; pp. 1–3. [CrossRef]
16. Tang, A.Y.; Drakinskiy, V.; Yhland, K.; Stenarson, J.; Bryllert, T.; Stake, J. Analytical Extraction of a Schottky Diode Model from Broadband S-Parameters. *IEEE Trans. Microw. Theory Tech.* **2013**, *61*, 1870–1878. [CrossRef]
17. Zeljami, K.; Gutierrez, J.; Pascual, J.P.; Fernandez, T.; Tazon, A.; Boussouis, M. Characterization and Modeling of Schottky Diodes up to 110 GHz for Use in Both Flip-chip and Wire-Bonded Assembled Environments. *Prog. Electromagn. Res.* **2012**, *131*, 457–475. [CrossRef]
18. Sze, S.M.; Ng, K.K. *Physics of Semiconductor Devices*; John Wiley & Sons Inc.: Hoboken, NJ, USA, 2007.
19. Advanced Design System (ADS). Available online: https://www.keysight.com/en/pc-1297113/advanced-design-system-ads?cc=ES&lc=eng (accessed on 14 May 2019).
20. Champlin, K.S.; Eisenstein, G. Cutoff frequency of submillimeter Schottky barrier diodes. *IEEE Trans. Microw. Theory Tech.* **1978**, *26*, 31–34. [CrossRef]

MDPI
St. Alban-Anlage 66
4052 Basel
Switzerland
Tel. +41 61 683 77 34
Fax +41 61 302 89 18
www.mdpi.com

Electronics Editorial Office
E-mail: electronics@mdpi.com
www.mdpi.com/journal/electronics

www.ingramcontent.com/pod-product-compliance
Lightning Source LLC
LaVergne TN
LVHW070841170726
843515LV00005B/544

* 9 7 8 3 0 3 6 5 0 9 9 6 9 *